Partial Update
Least-Square Adaptive Filtering

Synthesis Lectures on Communications

Editor
William Tranter, *Virginia Tech*

Code Division Multiple Access(CDMA)
R. Michael Buehrer
2006

Game Theory for Wireless Engineers
Allen B. MacKenzie and Luiz A. DaSilva
2006

Partial Update Least-Square Adaptive Filtering
Bei Xie and Tamal Bose

ISBN: 978-3-031-00553-4   paperback
ISBN: 978-3-031-01681-3   ebook

DOI 10.1007/978-3-031-01681-3

A Publication in the Springer series
SYNTHESIS LECTURES ON COMMUNICATIONS

Lecture #10
Series Editor: William Tranter, *Virginia Tech*
Series ISSN
Print 1932-1244   Electronic 1932-1708

Partial Update
Least-Square Adaptive Filtering

Bei Xie

Tamal Bose
University of Arizona

SYNTHESIS LECTURES ON COMMUNICATIONS #10

ABSTRACT

Adaptive filters play an important role in the fields related to digital signal processing and communication, such as system identification, noise cancellation, channel equalization, and beamforming. In practical applications, the computational complexity of an adaptive filter is an important consideration. The Least Mean Square (LMS) algorithm is widely used because of its low computational complexity ($O(N)$) and simplicity in implementation. The least squares algorithms, such as Recursive Least Squares (RLS), Conjugate Gradient (CG), and Euclidean Direction Search (EDS), can converge faster and have lower steady-state mean square error (MSE) than LMS. However, their high computational complexity ($O(N^2)$) makes them unsuitable for many real-time applications. A well-known approach to controlling computational complexity is applying partial update (PU) method to adaptive filters. A partial update method can reduce the adaptive algorithm complexity by updating part of the weight vector instead of the entire vector or by updating part of the time. In the literature, there are only a few analyses of these partial update adaptive filter algorithms. Most analyses are based on partial update LMS and its variants. Only a few papers have addressed partial update RLS and Affine Projection (AP). Therefore, analyses for PU least-squares adaptive filter algorithms are necessary and meaningful.

This monograph mostly focuses on the analyses of the partial update least-squares adaptive filter algorithms. Basic partial update methods are applied to adaptive filter algorithms including Least Squares CMA (LSCMA), EDS, and CG. The PU methods are also applied to CMA1-2 and NCMA to compare with the performance of the LSCMA. Mathematical derivation and performance analysis are provided including convergence condition, steady-state mean and mean-square performance for a time-invariant system. The steady-state mean and mean-square performance are also presented for a time-varying system. Computational complexity is calculated for each adaptive filter algorithm. Numerical examples are shown to compare the computational complexity of the PU adaptive filters with the full-update filters. Computer simulation examples, including system identification and channel equalization, are used to demonstrate the mathematical analysis and show the performance of PU adaptive filter algorithms. They also show the convergence performance of PU adaptive filters. The performance is compared between the original adaptive filter algorithms and different partial-update methods. The performance is also compared among similar PU least-squares adaptive filter algorithms, such as PU RLS, PU CG, and PU EDS. In addition to the generic applications of system identification and channel equalization, two special applications of using partial update adaptive filters are also presented. One application uses PU adaptive filters to detect Global System for Mobile Communication (GSM) signals in a local GSM system using the Open Base Transceiver Station (OpenBTS) and Asterisk Private Branch Exchange (PBX). The other application uses PU adaptive filters to do image compression in a system combining hyperspectral image compression and classification.

KEYWORDS

partial update, adaptive filter, LSCMA, RLS, EDS, CG

Contents

Acknowledgments

We would like to give special thanks to Dr. Brian Agee, who gave us a lot of advice and help on the CMA and LSCMA work. We would also like to thank Dr. Erzébet Merényi who helped us in the work on hyperspectral image compression and classification. Our thanks also go to Dr. Majid Manteghi and Thomas Tsou, who helped us with the OpenBTS project.

Bei Xie and Tamal Bose
May 2014

CHAPTER 1

Introduction

1.1 MOTIVATION

Adaptive filters play an important role in the fields related to digital signal processing and communication, such as system identification, noise cancellation, channel equalization, and beamforming. In practical applications, the computational complexity of an adaptive filter is an important consideration. The Least Mean Square (LMS) algorithm is widely used [21] because of its low computational complexity ($O(N)$) and simplicity in implementation. However, it is well known that the LMS has low convergence speed, especially for correlated input signals. The least squares algorithms, such as Recursive Least Squares (RLS), Conjugate Gradient (CG), and Euclidean Direction Search (EDS), can converge fast and have low steady-state mean square error (MSE). However, with high computational complexity ($O(N^2)$), these algorithms need expensive real-time resources, i.e., clock cycles, memory, and power in a digital signal processor (DSP) or field-programmable gate array (FPGA). A well-known approach to controlling computational complexity is applying partial-update (PU) method to adaptive filters. A partial-update adaptive filter reduces computational complexity by updating part of the coefficient vector instead of updating the entire vector or by updating part of the time. Moreover, the partial-update adaptive filters may converge faster than the full-update filters and achieve lower steady-state MSE in particular applications [11]. In the literature, partial-update methods have been applied to several adaptive filters, such as LMS, NLMS, RLS, Affine Projection (AP), Normalized Constant Modulus Algorithm (NCMA), etc. However, there are only a few analyses of these partial-update adaptive filter algorithms. Most analyses are based on partial-update LMS and its variants. Only a few papers have addressed partial-update RLS and AP.

1.2 PROBLEM STATEMENT

In this monograph, the basic partial-update methods are applied to adaptive filters which have high computational complexity. We have chosen two least-squares (LS) algorithms, CG and EDS. Both of these can converge fast and achieve small steady-state MSE. The complexity is less than the RLS algorithm, but is still $O(N^2)$. The PU methods are also applied to the Least Squares CMA (LSCMA). The LSCMA converges much faster than the CMA1-2 and NCMA. Performance of CMA-2 and NCMA are also analyzed in order to compare with the performance of the LSMCA. The computational complexity $O(N^2)$ per sample has limited the LSCMA to applications with a small number of adapted elements such as adaptive arrays and polarization combiners. The PU LSCMA can reduce the computational complexity and extend the LSCMA in applications with long filters and fractionally-spaced equalizers (FSEs).

Mathematical analysis has been done for the new partial-update adaptive filter algorithms. Mathematical analysis has also been extended to the existing PU adaptive filter algorithms. This work has analyzed the convergence conditions, steady-state performance, and tracking performance. The performance is compared between the original adaptive filter algorithms and different partial-update methods. Since a specific PU method in one adaptive filter algorithm which achieves good performance may not perform well in another adaptive filter algorithm, the performance of one PU method for different adaptive filter algorithms is also compared. Computational complexity is calculated for each partial-update method and each adaptive filter algorithm.

The major contributions of this work are summarized as follows.

1. Basic partial-update methods are applied to adaptive filter algorithms including CMA1-2, NCMA, Least Squares CMA (LSCMA), EDS, and CG.

2. Mathematical derivation and performance analysis are provided including convergence conditions, steady-state mean, and mean-square performance for a time-invariant system. The steady-state mean and mean-square performance are also presented for a time-varying system.

3. Computational complexity is calculated for each adaptive filter algorithm. It is also compared for each partial-update method. Numerical examples are presented to show how many multiplications or additions can be saved by using a PU adaptive filter.

4. Proper computer simulations demonstrate the mathematical analysis and show the performance of PU adaptive filter algorithms. The applications include system identification and channel equalization. The computer simulations also show the convergence performance of PU adaptive filters. The performance is compared between the original adaptive filter algorithms and different partial-update methods. The PU adaptive filters usually can achieve comparable performance to the full-update filters while reducing the computational complexity significantly.

5. Besides applications of system identification and channel equalization, two special applications of using partial update adaptive filters are also presented. One application uses PU adaptive filters to detect Global System for Mobile Communication (GSM) signals in a local GSM system using OpenBTS and Asterisk PBX. The other application is using PU adaptive filters to do image compression in a system combining hyperspectral image compression and classification.

1.3   ORGANIZATION OF THE MONOGRAPH

This monograph is organized as follows. Chapter 2 introduces basic adaptive signal processing models, partial-update methods, and a literature review on existing partial-update adaptive filter algorithms. Chapter 3 shows the performance of partial-update CMA-based algorithms, including CMA1-2, NCMA, and LSCMA. Chapters 4 and 5 show the performance of the partial-update CG and partial-update EDS algorithms, respectively. The comparison among PU RLS, CG, and EDS is shown. Chapter 6 shows the special applications of partial-update adaptive filters.

CHAPTER 2

Background

2.1 BASIC ADAPTIVE FILTER MODELS

Adaptive filters have been used in many applications, such as in system identification, noise cancellation, echo cancellation, channel equalization, signal prediction, and beamforming. In this section, models of system identification and channel equalization will be described in detail because they will be used as applications in Chapters 3 to 5. The basic idea of the adaptive filter consists of calculating the filter weights, calculating the estimated signal from the input signal and filter weights, taking the error between the estimated signal and the desired signal, and using the error to update the filter weights by some adaptive algorithm.

2.2 ADAPTIVE FILTER MODELS

The common model of an adaptive filter is shown in Fig. 2.1. In this model, $d(n)$ is the desired

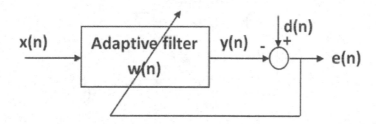

Figure 2.1: Adaptive filter system model.

signal, $x(n)$ is the input signal, $y(n)$ is the estimated signal, and $e(n)$ is the error between the desired signal and the estimated signal, that is $e(n) = d(n) - y(n)$. Let $\mathbf{w}(n)$ be the weight vector of the filter with length N. Define the input sequence vector as $\mathbf{x}(n) = [x(n), x(n-1), ..., x(n-N+1)]^T$. Therefore the estimated signal is $y(n) = \mathbf{w}^H(n)\mathbf{x}(n)$. The general algorithm can be summarized as follows:

$$y(n) = \mathbf{w}^H(n)\mathbf{x}(n), \tag{2.1}$$
$$e(n) = d(n) - y(n), \tag{2.2}$$
$$w(n+1) = w(n) + F\{e(n)\}, \tag{2.3}$$

where $F\{e(n)\}$ is a function of error term. The adaptive algorithms are usually designed to minimize some cost function such as the expectation of the instantaneous squared error $J(n) = E\{e^2(n)\}$ or the sum of squared errors up to the present sample $J(n) = \sum_{i=0}^{n} e^2(i)$. The adaptive filter usually takes a certain time to achieve the minimum of the cost function. This is referred to as the convergence rate.

2.2.1 SYSTEM IDENTIFICATION

System identification is used to estimate an unknown linear or nonlinear system in digital signal processing. A block diagram for system identification is given in Fig. 2.2. An input is applied to the unknown system and the adaptive filter simultaneously. Usually a noise will be added at the output of the unknown system. If the unknown system does not change with time, then it is a time-invariant system. If the unknown system changes with time, then it is a time-varying system.

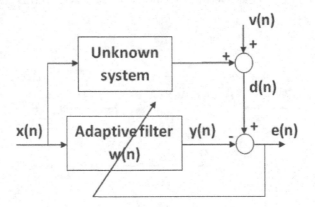

Figure 2.2: System identification model.

The system identification model can be presented as:

$$d(n) = \mathbf{x}^T(n)\mathbf{w}^o + v(n), \tag{2.4}$$

where $d(n)$ is the desired signal, $\mathbf{x}(n) = [x(n), x(n-1), ..., x(n-N+1)]^T$ is the input data vector of the unknown system, $\mathbf{w}^o(n) = [w_1^o, w_2^o, ..., w_N^o]^T$ is the impulse response vector of the unknown system, and $v(n)$ is zero-mean white noise, which is independent of any other signals.

Let $\mathbf{w}$ be the coefficient vector of an adaptive filter. The estimated signal $y(n)$ is defined as

$$y(n) = \mathbf{x}^T(n)\mathbf{w}(n), \tag{2.5}$$

and the output signal error is defined as

$$e(n) = d(n) - \mathbf{x}^T(n)\mathbf{w}(n). \tag{2.6}$$

The error will be fed back to update the adaptive filter.

2.2.2 CHANNEL EQUALIZATION

In a typical wireless communication system, the transmitted signal is subjected to noise and multipath which causes delay, distortion, and inter-symbol-interference (ISI). A channel equalizer is generally used to remove the multipath channel effects. A typical wireless communication system is shown in Fig. 2.3. The channel and equalizer shown in Fig. 2.3 can be modeled as FIR filters, which are shown in Fig. 2.4.

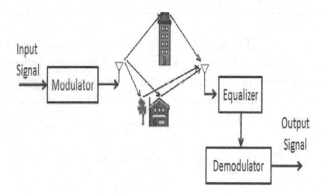

Figure 2.3: Channel equalization model.

Using the transmission channel shown in Fig. 2.4, the received signal $x(n)$ is modeled by

$$\begin{aligned} x(n) &= \varepsilon(n) + \sum_{\ell=0}^{L-1} c_\ell s(n-\ell) \\ &= \varepsilon(n) + \mathbf{s}^T(n)\mathbf{c}, \end{aligned} \tag{2.7}$$

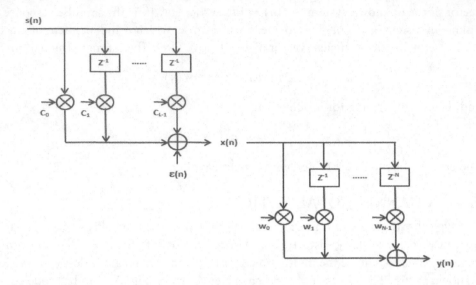

Figure 2.4: FIR channel and equalizer.

where $\mathbf{c} = [\ c_0\ \ \cdots\ \ c_{L-1}\ ]^T$ is the $L \times 1$ vector representation of the FIR channel impulse response, $\mathbf{s}(n) = [\ s(n)\ \ \cdots\ \ s(n - L + 1)\ ]^T$ is the $L \times 1$ transmitted signal state vector, and $\varepsilon(n)$ is the background noise and interference added to the desired signal at the receiver, and where $(\cdot)^T$ denotes the matrix transpose operation. Similarly, the recovered (equalizer output) signal is given by

$$
\begin{aligned}
y(n) &= \sum_{i=0}^{N-1} w_i x(n - i) \\
&= \mathbf{x}^T(n)\mathbf{w},
\end{aligned} \tag{2.8}
$$

where $\mathbf{w} = [\ w_0\ \ \cdots\ \ w_{N-1}\ ]^T$ is the $N \times 1$ vector representation of the FIR equalizer impulse response (referred to here as the *equalizer weight vector*), and where

$$
\mathbf{x}(n) = [\ x(n)\ \ \cdots\ \ x(n - N + 1)\ ]^T, \tag{2.9}
$$

is the $N \times 1$ received signal state vector. The goal of the equalizer adaptation algorithm is to design an equalizer weight vector that can remove the channel distortion from the desired signal, such that the convolved channelizer and equalizer impulse response $\sum c_m w_{n-m} \approx \delta_q$ for some group delay q, without unduly raising the signal-to-noise ratio (SNR) of the equalizer output signal. In conventional nonblind systems, this is typically accomplished by adapting the equalizer to minimize the mean-square error (MSE) between the equalizer output signal $y(n)$ and the transmitted signal $s(n)$ over some known component of $s(n)$, e.g., some known training or pilot sequence transmitted as part of $s(n)$. For a blind equalizer, a training sequence is not used.

In practice, a decision-directed channel equalization is often used for a nonblind system. The decision-directed channel equalization diagram is shown in Fig. 2.5 [43]. There are two modes: training mode and decision-directed mode. In the training mode, a training signal is used to obtain an appropriate set of the adaptive filter coefficients. The error between the delayed desired signal and the output of the adaptive filter is used to update the adaptive filter coefficients. In the decision-directed mode, there is no training signal. The output the adaptive filter is mapped to the ideal symbol according to the shortest Euclidean distance. This signal represents the estimated desired signal. The error between the delayed version of the estimated desired signal and the output of the adaptive filter is fed back to update the adaptive filter.

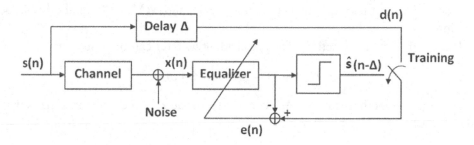

Figure 2.5: Decision-directed equalizer.

2.3   EXISTING WORK ON PARTIAL UPDATE ADAPTIVE FILTERS

The partial-update (PU) method is a straightforward approach to controlling the computational complexity because it only updates part of the coefficient vector instead of updating the entire filter vector. In the literature, partial update methods have been mostly studied for LMS and its variants [12], [13], [19], [34], [29], [54], [11]. Partial-update methods such as periodic PU, sequential PU, stochastic PU, MMax update method, selective PU method, set membership PU method, and the block PU method have been applied to LMS or NLMS. The stability condition, robustness, mean convergence analysis, mean-square convergence analysis, and tracking performance have been done for LMS or NLMS. Moreover, the PU LMS for sparse impulse responses [8], [30] and PU Transform-Domain LMS [1], [2] have been addressed. In [14, 40, 45, 46], new PU methods as such proportionate normalized least-mean-squares (PNLMS), improved PNLMS (IPNLMS), select and queue with a constraint (SELQUE) method, and modified SELQUE (M-SELQUE) method are also developed for sparse impulse responses and aim to improve the convergence rate of NLMS in echo cancellation systems. In [38], the MMax RLS and MMax AP have been analyzed for white inputs. In [29], the tracking performance has been analyzed for MMax RLS and MMax AP. In [6], an analysis of the transient and the steady-state behavior of a filtered-x partial-error affine projection algorithm was provided. In [53], [9], set-membership AP was proposed, but the mathematic analysis was not provided. In [10], [11], the selective PU NCMA has been proposed. However, the mathematic analysis was not provided.

2.4 BASIC PARTIAL UPDATE METHODS

Instead of updating all of the $N \times 1$ coefficients, the partial-update method only updates $M \times 1$ coefficients, where $M < N$. In this work, we consider the basic partial update methods including periodic PU [13], sequential PU [13], stochastic PU [19], and MMax update method [12]. The set-membership PU method is not considered here because it was developed for steepest-descent methods such as LMS and AP. It aims to adjust a time-variant step-size for these algorithms. A detailed description of periodic, sequential, stochastic, and MMax PU methods will be given in this section.

The weight-update function of a typical adaptive filter can be written as

$$\mathbf{w}(n + 1) = \mathbf{w}(n) + \Delta \mathbf{w}(n). \tag{2.10}$$

The partial-update method chooses M elements from the Δw and generates the new weights. It modifies (2.10) to

$$w_i(n + 1) = \begin{cases} w_i(n) + \Delta w_i(n) & \text{if } i \in \mathcal{I}_M(n) \\ w_i(n) & \text{otherwise} \end{cases}, \tag{2.11}$$

where w_i means the ith element of $\mathbf{w}$ and $\mathcal{I}_M(n)$ is a subset of $\{1, 2, \cdots, N\}$ with M elements at time n. For different PU methods, the subset $\mathcal{I}_M(n)$ is different at different time. The weight-update function of PU adaptive filters can be also expressed as

$$\mathbf{w}(n + 1) = \mathbf{w}(n) + \hat{\Delta} \mathbf{w}(n), \tag{2.12}$$

where

$$\hat{\Delta w_i}(n) = \begin{cases} \hat{\Delta w_i}(n) & \text{if } i \in \mathcal{I}_M(n) \\ 0 & \text{otherwise} \end{cases}. \tag{2.13}$$

2.4.1 PERIODIC PARTIAL UPDATE METHOD

The periodic partial update method only updates the coefficients at every S^{th} iteration and copies the coefficients at the other iterations, where $S = \lceil \frac{N}{M} \rceil$ which is the ceiling of $\frac{N}{M}$. The update function can be written as

$$\mathbf{w}(S(n + 1)) = \mathbf{w}(Sn) + \Delta \mathbf{w}(Sn). \tag{2.14}$$

Obviously, this method can reduce the overall computational cost. This method works well for a static channel. After filter convergence, there is no need to update the filter frequently since the channel does not change. Since periodic PU algorithms update the whole vector, the steady-state performance will be the same as the original adaptive filter algorithms for stationary input. The convergence rate of periodic PU algorithms will be S times slower than the original algorithms.

2.4.2   SEQUENTIAL PARTIAL UPDATE METHOD

The sequential partial update method designs the subset as

$$\mathcal{I}_M(n) = \mathcal{I}_M(n \mod S + 1),$$

so that

$$
\begin{aligned}
\mathcal{I}_M(1) &= \{1, 2, \cdots, M\}, \\
\mathcal{I}_M(2) &= \{M + 1, M + 2, \cdots, 2M\}, \\
&\vdots \\
\mathcal{I}_M(S) &= \{(S - 1)M + 1, (S - 1)M + 2, \cdots, N\}.
\end{aligned}
$$

The sequential PU method chooses the subset of the weights in a round-robin fashion. For example, there are three elements in the weight vector and we let the PU length be 1. The sequential PU method may choose the first element at iteration n, choose the second element at iteration $n + 1$, and the third element at iteration $n + 2$. Updating a subset of the weights can reduce computational complexity. With a round-robin fashion, each element of the filter will be updated evenly.

2.4.3   STOCHASTIC PARTIAL UPDATE METHOD

The stochastic method can be implemented as a randomized version of the sequential method [11]. It can choose the sequential subsets ($\mathcal{I}_M(1)$ - $\mathcal{I}_M(S)$) randomly. In this book, the stochastic method is implemented as M elements which are randomly chosen from $\{1, 2, \cdots, N\}$ rather than the sequential subsets which are randomly chosen. The uniformly distributed random process will be applied. Therefore, for each iteration, M of N weight elements will be chosen and updated with probability $p_i = 1/S$. The stochastic method may eliminate the instability problems of the sequential method for some non-stationary inputs [11].

2.4.4   MMAX METHOD

The subset becomes

$$
\begin{aligned}
\mathcal{I}_M(n) &= \{k\}, \\
k &\in \text{argmax}_{l \in \{1,2,\cdots,N\}}\{|\mathbf{x}_l(n)|, M\},
\end{aligned}
\tag{2.15}
$$

which means the elements of the weight vector $\mathbf{w}$ are updated according to the position of the first M-max elements of the input $\mathbf{x}$. This method aims to find the subset of the update vector that can make the biggest contribution to the convergence. In some applications, this method may converge faster than the full-update algorithm because the eigenspread of the partial-update input autocorrelation matrix may be smaller than the eigenspread of the full-update input autocorrelation matrix. Although choosing M-max elements of the input vector can increase the computational complexity, there are fast sorting algorithms [11] to reduce the computational cost.

CHAPTER 3

Partial Update CMA-based Algorithms for Adaptive Filtering

In this chapter, the partial-update methods are applied to CMA1-2, NCMA, and LSCMA. The theoretical mean and mean-square analyses of the PU CMA1-2, NCMA, and dynamic LSCMA for both time-invariant and time-varying systems are derived. The computational complexity is compared for the original CMAs and PU CMAs.

3.1 MOTIVATION

The slow convergence of conventional steepest-descent CMA (SDCMA) and high computational complexity of least-squares CMA (LSCMA) make the CMA unsuitable for applications with dynamic environments and rapidly-changing channel conditions. It is important to find a method for CMA which can increase the convergence rate while maintaining low computational complexity. With these properties, CMA algorithms can be applied to high-mobility systems where the channel changes rapidly over the signal reception interval, and packet-data and VoIP communication systems where the transmitted signal has very short duration or is subject to rapid changes in background interference.

3.2 REVIEW OF CONSTANT MODULUS ALGORITHMS

The CMA algorithm aims to remove the channel effect and recover the input signal modulus without using any training sequence. It has a nonconvex cost function [22]

$$J(n) = E\{(|y(n)|^p - 1)^2\}, \tag{3.1}$$

where p is a positive integer, and 1 is the modulus of some signals, such as PSK, QPSK (4-QAM), GMSK, and FM. In [48], the expectation was defined as a time-average expectation and the cost function was named the constant modulus cost function (CMCF).

Two cases of CMA are widely studied [22]. They are CMAp-2 with $p = 1$ and CMAp-2 with $p = 2$. In this chapter, CMA1-2 will be studied for partial update methods. The CMA1-2 algorithm is chosen because the cost functions of CMA1-2 (1-2 CMCF) and *normalized constant modulus algorithm* (NCMA) are the same, and the cost function of LSCMA is also related to the 1-2 CMCF.

The 1-2 CMCF [22] is

$$J_{\mathbf{w}}(n) = E\{(|y(n)| - 1)^2\}, \tag{3.2}$$

where $E\{\cdot\}$ denotes the time-averaging expectation. This cost function is equivalent to the least-squares cost function

$$J_{\mathbf{w}}(n) = E\{|y(n) - \hat{s}(n)|^2\}, \tag{3.3}$$

where $\hat{s}(n) = sgn(y(n)) = \frac{y(n)}{|y(n)|}$ is the complex sign of $y(n)$. The CMA reported in [18], [48] optimizes (3.2) by using a *stochastic gradient* approximation to the *steepest-descent* method to minimize $J(\mathbf{w})$,

$$
\begin{aligned}
\mathbf{w} \quad \leftarrow \quad & \mathbf{w} - \frac{\mu}{2} \nabla_{\mathbf{w}} J_{\mathbf{w}}(n) \\
= \quad & \mathbf{w} - \mu E\{\mathbf{x}^*(n)\,(y(n) - \hat{s}(n))\}, \\
= \quad & \mathbf{w} + \mu E\{\mathbf{x}^*(n)\,(\hat{s}(n) - y(n))\},
\end{aligned}
\tag{3.4}
$$

where μ is the step size, and $\nabla_{\mathbf{w}} J_{\mathbf{w}}(n)$ is the derivative of the cost function $J_{\mathbf{w}}(n)$ with respect to $\mathbf{w}$. The stochastic gradient approximation dispenses with the time-averaging operation given in (3.4), yielding the CMA1-2 reported in [18], [48],

$$\mathbf{w}(n+1) \quad = \quad \mathbf{w}(n) + \mu \mathbf{x}^*(n)\,(\hat{s}(n) - y(n)) \tag{3.5}$$

such that $\mathbf{w}$ is updated at every time sample n using an instantaneous measure of the gradient at that time sample. A variant of the 1-2 CMA referred to as the NCMA [24] is given by

$$\mathbf{w}(n+1) = \mathbf{w}(n) + \mu \frac{\mathbf{x}^*(n)}{\|\mathbf{x}(n)\|^2}\,(\hat{s}(n) - y(n)) \tag{3.6}$$

and has also been proposed to improve the stability of the CMA1-2. Both of these methods are referred to here as *steepest-descent CMAs* (SDCMAs), due to their close relationship to (3.4).

LSCMA was originally proposed to find the optimal beamforming weights which can minimize the least squares cost function (3.3). Since it updates the filter weights using a block-update method, the cost function for a block of data with length K can be written as

$$\|\mathbf{y} - \hat{\mathbf{s}}\|^2, \tag{3.7}$$

where $\mathbf{y}$ is the recovered signal vector, and $\hat{\mathbf{s}}$ is the normalized recovered signal vector. Define x as the received signal, $\mathbf{w}$ as the blind adaptive filter with length N, $y = \mathbf{x}^T \mathbf{w}$ as the recovered signal, and $\hat{s} = \frac{y}{|y|}$ as the normalized recovered signal. Represent the signal waveforms $\{\mathbf{x}_i\}_{i=1}^K$, $\{y_i\}_{i=1}^K$, and $\{\hat{s}_i\}_{i=1}^K$ in matrix form

$$\mathbf{X}(n) = [\mathbf{x}(n), \mathbf{x}(n-1), ..., \mathbf{x}(n-K+1)]^T, \tag{3.8}$$

where $\mathbf{x}(n) = [x(n), x(n-1), ..., x(n-N+1)]^T$ and $\mathbf{X}$ is a matrix with size $K \times N$ ($K > N$),

$$\begin{aligned} \mathbf{y}(n) &= [y(n), y(n-1), ..., y(n-K+1)]^T \\ &= \mathbf{X}(n)\mathbf{w}(n), \end{aligned} \tag{3.9}$$

$$\hat{\mathbf{s}}(n) = \left[\frac{y_1(n)}{|y_1(n)|}, \frac{y_2(n)}{|y_2(n)|}, ..., \frac{y_K(n)}{|y_K(n)|} \right]. \tag{3.10}$$

The update equation for LSCMA [3] is

$$\begin{aligned} \mathbf{w}(n+1) &= \mathbf{X}^\dagger(n)\hat{\mathbf{s}}(n), \tag{3.11} \\ &= \mathbf{w}(n) + \mathbf{X}^\dagger(n)(\hat{\mathbf{s}}(n) - \mathbf{y}(n)), \tag{3.12} \end{aligned}$$

where $(.)^\dagger$ denotes the Moore-Penrose pseudoinverse. For full rank $\mathbf{X}$, $\mathbf{X}^\dagger = (\mathbf{X}^H \mathbf{X})^{-1} \mathbf{X}^H$.

Both the static LSCMA and the dynamic LSCMA update the weights block-by-block. However, the static LSCMA may update the weights several times by using the same block of data, while the dynamic LSCMA updates the weights only once for each block. Obviously, the dynamic LSCMA can reduce the computational cost when compared with the static LSCMA. It can also achieve fast convergence for a dynamic environment [3].

3.3   PARTIAL UPDATE CONSTANT MODULUS ALGORITHMS

In this section, partial-update methods are applied to CMA1-2, Normalized CMA (NCMA), and LSCMA.

3.3.1   PARTIAL UPDATE CMA

The partial-update methods introduced in Section II are applied to the CMA1-2 algorithm. The weight-update equation of PU CMA1-2 is

$$\mathbf{w}(n+1) = \mathbf{w}(n) + \mu \hat{\mathbf{x}}(n)^* \left(\frac{y(n)}{|y(n)|} - y(n) \right), \tag{3.13}$$

where $\hat{\mathbf{x}}$ is the partial update data vector and is defined as

$$\hat{x}_i(n) = \begin{cases} x_i(n) & \text{if } i \in \mathcal{I}_M(n) \\ 0 & \text{otherwise} \end{cases}, \tag{3.14}$$

and $y(n) = \mathbf{x}^T(n)\mathbf{w}(n)$. The partial update subset $\mathcal{I}_M(n)$ is defined in Section 2.4.

3.3.2 PARTIAL UPDATE NCMA

The weight-update equation of partial update NCMA is

$$\mathbf{w}(n+1) = \mathbf{w}(n) + \mu \frac{\hat{\mathbf{x}}^*(n)}{\|\hat{\mathbf{x}}(n)\|^2} \left(\frac{y(n)}{|y(n)|} - y(n) \right). \tag{3.15}$$

In [11], Equation (3.15) with subset (2.15) was named as selective-partial-update NCMA, not MMax NCMA. However, we name it MMax NCMA to be consistent with other algorithms.

3.3.3 PARTIAL UPDATE LSCMA

The partial update LSCMA has the uniform update equation:

$$\mathbf{w}(q+1) = \hat{\mathbf{X}}^\dagger(q)\hat{\mathbf{s}}(q) \tag{3.16}$$
$$= \mathbf{w}(q) + \hat{\mathbf{X}}^\dagger(q)(\hat{\mathbf{s}}(q) - \mathbf{y}(q)), \tag{3.17}$$

where $\mathbf{y}(q)$ still equals $\mathbf{X}(q)\mathbf{w}(q)$, and $\hat{\mathbf{X}}(q)$ is a $K \times N$ sparse matrix of the partial-update equalizer state vectors selected over block q,

$$\hat{\mathbf{X}}(q) = \begin{bmatrix} \hat{\mathbf{x}}_0(q) & \cdots & \hat{\mathbf{x}}_{N-1}(q) \end{bmatrix}, \tag{3.18}$$

where

$$\hat{\mathbf{x}}_i(q) = \begin{cases} \mathbf{x}_i(q) & \text{if } i \in \mathcal{I}_M(q) \\ \mathbf{0} & \text{otherwise} \end{cases}, \tag{3.19}$$

where $\mathbf{x}_i(q)$ is defined as the ith column of the matrix $\mathbf{X}$ and $\hat{\mathbf{x}}_i(q)$ is the ith column of the matrix $\hat{\mathbf{X}}$. Only M columns of $\mathbf{X}$ are used to update the equation. For different PU methods, the subset $\mathcal{I}_M(q)$ will be different. However, the subset of MMax LSCMA is different from (2.15). It becomes

$$\mathcal{I}_M(q) = \{k\},$$
$$k \in \text{argmax}_{l \in \{1,2,\cdots,N\}}\{\|\mathbf{x}_l(q)\|, M\}, \tag{3.20}$$

which means the elements of the weight vector $\mathbf{w}$ are updated in the positions of the M-max columns of the matrix $\mathbf{X}$ with respect to the Euclidean norm. Examination of (3.18) reveals a simpler alternate form for (3.17),

$$\mathbf{w}(q+1) = \mathbf{w}(q) + \mathbf{T}(\mathbf{X}(q)\mathbf{T}(q))^\dagger (\hat{\mathbf{s}}(q) - \mathbf{y}(q)), \tag{3.21}$$

where $\mathbf{T}(q)$ is a *selection matrix* given by

$$\mathbf{T}(q) = \begin{bmatrix} \mathbf{e}_{i_1(q)} & \cdots & \mathbf{e}_{i_M(q)} \end{bmatrix}, \tag{3.22}$$

and where $\mathbf{e}_i$ is the ith $N \times 1$ Euclidean basis vector. The individual elements of $\mathbf{w}(q)$ are therefore updated using the partial-update formula

$$w_i(q+1) = \begin{cases} \tilde{w}_m(q+1), & i = i_m(q) \\ w_i(q), & \text{otherwise} \end{cases} \qquad (3.23)$$

$$\tilde{\mathbf{w}}(q+1) = \tilde{\mathbf{w}}(q) + \tilde{\mathbf{X}}^{\dagger}(q)(\hat{\mathbf{s}}(q) - \mathbf{y}(q)), \qquad (3.24)$$

where $\tilde{\mathbf{w}}(q+1)$ is the $M \times 1$ vector of the updated coefficients and $\tilde{\mathbf{X}}(q)$ is the $K \times M$ *pruned* matrix of the partial-update equalizer state vectors selected over block q,

$$\tilde{\mathbf{X}}(q) = \begin{bmatrix} \hat{\mathbf{x}}_{i_1(q)}(q) & \cdots & \hat{\mathbf{x}}_{i_M(q)}(q) \end{bmatrix}, \qquad (3.25)$$

i.e., with the zero-filled elements of $\hat{\mathbf{X}}(q)$ removed. $\tilde{\mathbf{X}}(q)$ has full rank $M < K$ in noisy environments, allowing computation of (3.24) using straightforward linear algebraic methods, e.g., QR decomposition. Examination of (3.21) shows that only M elements of $\mathbf{w}(q)$ are actually updated at each time instance. However, all $N \times 1$ weight elements must still be used to compute $\mathbf{y}(q)$ and $\hat{\mathbf{s}}(q)$.

3.4 ALGORITHM ANALYSIS FOR A TIME-INVARIANT SYSTEM

In this section, the theoretical mean and mean-square analyses of the PU CMA1-2, NCMA, and dynamic LSCMA are derived for a time-invariant system. The convergence analysis of static LSCMA is derived. The computational complexity is compared for the original CMAs and PU CMAs.

3.4.1 STEADY-STATE PERFORMANCE OF PARTIAL UPDATE SDCMA

Use the channel equalization system shown in Fig. 2.3 and Fig. 2.4. The mean-square error (MSE) can be defined as

$$E\{|e(n)|^2\} = E\{|y(n) - s(n - \Delta)|^2\}, \qquad (3.26)$$

where $y(n) = \mathbf{x}^T(n)\mathbf{w}(n)$, $s(n)$ is the transmitted signal, and $s(n - \Delta)$ is the delayed version of the transmitted signal. Assume that there is an optimal weight $\mathbf{w}^o$ that satisfies $(\mathbf{x}(n) - (n))^T\mathbf{w}^o = s(n - \Delta)$, where $\mathbf{x}(n)$ is the received signal, and (n) is channel noise. The phase rotation is not considered here.

The MSE can be rewritten as

$$E\{|e(n)|^2\} = E\{|\mathbf{x}^T(n)\mathbf{w}(n) - \mathbf{x}^T(n)\mathbf{w}^o + v(n)|^2\}, \qquad (3.27)$$

where $v(n) = {}^T(n)\mathbf{w}^o$ is the noise component of the optimal equalizer output signal and $(n) = [\varepsilon(n - i)]_{i=0}^{N-1}$ is the noise component of the equalizer state vector. If $\varepsilon(n)$ is i.i.d. with mean zero and is independent from the desired signal $s(n)$, then $v(n)$ is zero mean and is also independent from the desired signal.

Introduce the coefficient error vector

$$\mathbf{z}(n) = \mathbf{w}^o - \mathbf{w}(n). \tag{3.28}$$

To determine the steady-state mean-square behavior of the partial update CMA1-2, two assumptions are needed.

Assumption I: The coefficient error $\mathbf{z}(n)$ is independent of the input signal $\mathbf{x}(n)$ in steady state.

Assumption II: Noise $v(n)$ is independent of the input signal $\mathbf{x}(n)$ and has zero mean.

The MSE becomes

$$E\{|e(n)|^2\} = \sigma_v^2(n) + E\{\left|\mathbf{x}^T(n)\mathbf{z}(n)\right|^2\}. \tag{3.29}$$

The PU CMA1-2 coefficient error recursion equation is obtained by subtracting both sides of (3.13) from $\mathbf{w}^o$,

$$\hat{\mathbf{z}}(n+1) = \hat{\mathbf{z}}(n) - \mu\hat{\mathbf{x}}^*(n)(\frac{y(n)}{|y(n)|} - y(n)), \tag{3.30}$$

where each element of $\hat{\mathbf{z}}(n)$ is defined as

$$\hat{z}_i(n) = \begin{cases} z_i(n) & \text{if } i \in \mathcal{I}_M(n) \\ 0 & \text{otherwise} \end{cases}. \tag{3.31}$$

Define

$$\begin{aligned} e(n) &= \frac{y(n)}{|y(n)|} - y(n) \\ &= \mathbf{x}^T(n)\mathbf{z}(n) - v(n). \end{aligned} \tag{3.32}$$

We can rewrite (3.30) as

$$\begin{aligned} \hat{\mathbf{z}}(n+1) &= \hat{\mathbf{z}}(n) - \mu\hat{\mathbf{x}}^*(n)e(n) \\ &= \hat{\mathbf{z}}(n) + \mu\hat{\mathbf{x}}^*(n)v(n) - \mu\hat{\mathbf{x}}^*(n)\mathbf{x}^T(n)\mathbf{z}(n). \end{aligned} \tag{3.33}$$

Taking the expectation of (3.33), we get

$$E\{\hat{\mathbf{z}}(n+1)\} = E\{\hat{\mathbf{z}}(n)\} + \mu E\{\hat{\mathbf{x}}^*(n)v(n)\} - \mu E\{\hat{\mathbf{x}}^*(n)\mathbf{x}^T(n)\mathbf{z}(n)\}. \tag{3.34}$$

For steady state, $E\{\hat{\mathbf{z}}(n+1)\} = E\{\hat{\mathbf{z}}(n)\}$. Using independence assumptions, we obtain

$$E\{\hat{\mathbf{x}}^*\mathbf{x}^T\}E\{\mathbf{z}\} = E\{\hat{\mathbf{x}}^*\}E\{v\}. \tag{3.35}$$

Since $E\{v\} = 0$, $E\{\mathbf{z}\} = \mathbf{0}$, the average weights converge to the optimal weights at steady state. The convergence rate of PU CMA1-2 depends on the term $\mu E\{\hat{\mathbf{x}}^*\mathbf{x}^T\}$. Notice

$$E\{\hat{\mathbf{x}}^*\mathbf{x}^T\} = \beta E\{\mathbf{x}^*\mathbf{x}^T\}, \tag{3.36}$$

where

$$\begin{cases} \beta = \frac{M}{N}, & \text{for sequential and stochastic methods} \\ \frac{M}{N} < \beta < 1, & \text{for MMax method} \end{cases}, \tag{3.37}$$

if $s(n)$ is stationary. For the same step size μ, the convergence rate of sequential and stochastic CMA1-2 is N/M times smaller than the original CMA1-2. For the MMax method, the convergence rate is greater than the sequential and stochastic CMA1-2.

Convergence in the mean is achieved if

$$0 < \mu < \frac{2}{\beta \lambda_{max}(E\{\mathbf{x}^*\mathbf{x}^T\})}, \tag{3.38}$$

where $\lambda_{max}(.)$ means the maximal eigenvalue of $(.)$ and β satisfies (3.37).

Multiplying $\hat{\mathbf{x}}^T(n)$ to both sides of (3.33), we have

$$\hat{\mathbf{x}}^T(n)\hat{\mathbf{z}}(n+1) = \hat{\mathbf{x}}^T(n)\hat{\mathbf{z}}(n) - \mu\|\hat{\mathbf{x}}(n)\|^2\mathbf{x}^T(n)\mathbf{z}(n) + \mu\|\hat{\mathbf{x}}(n)\|^2 v(n). \tag{3.39}$$

Multiplying $\hat{\mathbf{z}}^H(n+1)\hat{\mathbf{x}}^*(n)/\|\hat{\mathbf{x}}(n)\|^2$ by $\hat{\mathbf{x}}^T(n)\hat{\mathbf{z}}(n+1)$ and using (3.39), we get

$$\begin{aligned}
&\frac{\hat{\mathbf{z}}^H(n+1)\hat{\mathbf{x}}^*(n)}{\|\hat{\mathbf{x}}(n)\|^2} \times \hat{\mathbf{x}}^T(n)\hat{\mathbf{z}}(n+1) \\
&= \frac{\hat{\mathbf{z}}^H(n+1)\hat{\mathbf{x}}^*(n)\hat{\mathbf{x}}^T(n)\hat{\mathbf{z}}(n+1)}{\|\hat{\mathbf{x}}(n)\|^2} \\
&- \mu\hat{\mathbf{z}}^H(n)\hat{\mathbf{x}}^*(n)\mathbf{x}^T(n)\mathbf{z}(n) - \mu\mathbf{z}^H(n)\mathbf{x}^*(n)\hat{\mathbf{x}}^T(n)\hat{\mathbf{z}}(n) \\
&+ \mu^2\mathbf{z}^H(n)\mathbf{x}^*(n)\|\hat{\mathbf{x}}(n)\|^2\mathbf{x}^T(n)\mathbf{z}(n) \\
&+ \mu^2 v^*(n)\|\hat{\mathbf{x}}(n)\|^2 v(n) \\
&+ \mu\mathbf{z}^H(n)\hat{\mathbf{x}}^*(n)v(n) + \mu v^*(n)\hat{\mathbf{x}}^T(n)\mathbf{z}(n) \\
&- \mu^2\mathbf{z}^H(n)\mathbf{x}^*(n)\|\hat{\mathbf{x}}(n)\|^2 v(n) - \mu^2 v^*(n)\|\hat{\mathbf{x}}(n)\|^2\mathbf{x}^T\mathbf{z}(n). \tag{3.40}
\end{aligned}$$

Taking the expectation on both sides of (3.40) and using the independence assumption, we obtain $E\{\hat{\mathbf{z}}^H(n+1)\hat{\mathbf{x}}^*(n)\hat{\mathbf{x}}^T(n)\hat{\mathbf{z}}(n+1)/\|\hat{\mathbf{x}}(n)\|^2\} = E\{\hat{\mathbf{z}}^H(n)\hat{\mathbf{x}}^*(n)\hat{\mathbf{x}}^T(n)\hat{\mathbf{z}}(n)/\|\hat{\mathbf{x}}(n)\|^2\}$. In steady state, there is also $E\{\hat{\mathbf{z}}^H\hat{\mathbf{x}}^*\mathbf{x}^T\mathbf{z}\} = \beta E\{\mathbf{z}^H\mathbf{x}^*\mathbf{x}^T\mathbf{z}\}$, where β satisfies (3.37). Using independence assumption, $E\{v(n)\} = 0$, and after simplification, we obtain

$$2\beta E\{|\mathbf{x}^T\mathbf{z}|^2\} = \mu E\{\|\hat{\mathbf{x}}\|^2\}E\{|\mathbf{x}^T\mathbf{z}|^2\} + \mu E\{|v|^2\}E\{\|\hat{\mathbf{x}}\|^2\}. \tag{3.41}$$

Therefore, $E\{|\mathbf{x}^T\mathbf{z}|^2\}$ becomes

$$E\{|\mathbf{x}^T\mathbf{z}|^2\} = \frac{\mu\sigma_v^2 E\{\|\hat{\mathbf{x}}\|^2\}}{2\beta - \mu E\{\|\hat{\mathbf{x}}\|^2\}}. \tag{3.42}$$

The steady-state MSE of PU CMA1-2 becomes

$$E\{|e(n)|^2\} = \sigma_v^2 + \frac{\mu\sigma_v^2 E\{\|\hat{\mathbf{x}}\|^2\}}{2\beta - \mu E\{\|\hat{\mathbf{x}}\|^2\}}, \qquad n \to \infty. \tag{3.43}$$

Since $E\{\|\hat{\mathbf{x}}\|^2\} = \beta E\{\|\mathbf{x}\|^2\}$, the steady-state MSE of PU CMA1-2 is also equal to

$$E\{|e(n)|^2\} = \sigma_v^2 + \frac{\mu\sigma_v^2 E\{\|\mathbf{x}\|^2\}}{2 - \mu E\{\|\mathbf{x}\|^2\}}, \qquad n \to \infty, \tag{3.44}$$

which is exactly the same as the original CMA1-2.

Using the same analysis method, the weights of PU NCMA also converge to the optimal weights in steady state. The convergence rate of sequential, stochastic, and MMax NCMA depends on the term $E\{\frac{\hat{\mathbf{x}}\hat{\mathbf{x}}^H}{\|\hat{\mathbf{x}}\|^2}\}$, which is close to the term $E\{\frac{\mathbf{x}\mathbf{x}^H}{\|\mathbf{x}\|^2}\}$ of NCMA. The convergence rate of sequential, stochastic, and MMax NCMA is therefore close to that of the original NCMA. The mean stability is achieved if $0 < \mu < 2$. The steady-state MSE of PU NCMA becomes

$$E\{|e(n)|^2\} = \sigma_v^2 + \frac{\mu\sigma_v^2}{2 - \mu}, \qquad n \to \infty. \tag{3.45}$$

It is still the same as the original NCMA.

3.4.2 STEADY-STATE PERFORMANCE OF PARTIAL UPDATE DYNAMIC LSCMA

The dynamic LSCMA updates the weights only once for each block. The MSE of the whole sequence is

$$E\{\|\mathbf{y}(q) - \mathbf{s}(q)\|^2\}, \tag{3.46}$$

where the expectation is a time-averaging expectation. It can be also viewed as $E\{\|\mathbf{X}(q)\mathbf{w}(q) - \mathbf{X}(q)\mathbf{w}^* + \mathbf{v}(q)\|^2\}$, where $\mathbf{w}^o$ is the optimal weight and v is noise with zero-mean and variance σ_v^2, which is independent of any other signals.

The assumptions from the CMA1-2 analysis are still needed. There is another assumption needed for LSCMA analysis.

Assumption III: In steady state, the coefficient error $\mathbf{z}$ is small enough, and $\mathbf{Xz}$ is independent of the data matrix $\hat{\mathbf{X}}$.

The MSE becomes

$$E\{|e(q)|^2\} = K\sigma_v^2(q) + E\{\|\mathbf{X}(q)\mathbf{z}(q)\|^2\}. \tag{3.47}$$

Subtract both sides of (3.17) from $\mathbf{w}^o$, and we have

$$\hat{\mathbf{z}}(q+1) = \hat{\mathbf{z}}(q) - \hat{\mathbf{X}}^{\dagger}(q)(\hat{\mathbf{s}}(q) - \mathbf{y}(q)). \tag{3.48}$$

Define

$$
\begin{aligned}
\mathbf{e}(q) &= \hat{\mathbf{s}}(q) - \mathbf{y}(q) \\
&= \mathbf{X}(q)\mathbf{z}(q) - \mathbf{v}(q).
\end{aligned}
$$

(3.49)

We can rewrite the PU LSCMA coefficient error recursion equation as

$$
\begin{aligned}
\hat{\mathbf{z}}(q+1) &= \hat{\mathbf{z}}(q) - \hat{\mathbf{X}}^{\dagger}(q)\mathbf{e}(q) \\
&= \hat{\mathbf{z}}(q) + \hat{\mathbf{X}}^{\dagger}(q)\mathbf{v}(q) - \hat{\mathbf{X}}^{\dagger}(q)\mathbf{X}(q)\mathbf{z}(q).
\end{aligned}
$$

(3.50)

Taking the expectation on both sides of (3.50), we get

$$
E\{\hat{\mathbf{z}}(q+1)\} = E\{\hat{\mathbf{z}}(q)\} + E\{\hat{\mathbf{X}}^{\dagger}(q)\mathbf{v}(q)\} - E\{\hat{\mathbf{X}}^{\dagger}(q)\mathbf{X}(q)\mathbf{z}(q)\}.
$$

(3.51)

For steady state, $E\{\hat{\mathbf{z}}(q+1)\} = E\{\hat{\mathbf{z}}(q)\}$. Using independence assumptions, we obtain

$$
E\{\hat{\mathbf{X}}^{\dagger}\mathbf{X}\}E\{\mathbf{z}\} = E\{\hat{\mathbf{X}}^{\dagger}\}E\{\mathbf{v}\}.
$$

(3.52)

Since $E\{\mathbf{v}\} = \mathbf{0}$, $E\{\mathbf{z}\} = \mathbf{0}$, the average weights converge to the optimal weights at steady state. The convergence condition of PU LSCMA is that $E\{\hat{\mathbf{X}}^{\dagger}\mathbf{X}\}$ exists.

Multiply $\hat{\mathbf{X}}(q)$ to both sides of (3.50), and we have

$$
\hat{\mathbf{X}}(q)\hat{\mathbf{z}}(q+1) = \hat{\mathbf{X}}(q)\hat{\mathbf{z}}(q) - \hat{\mathbf{X}}(q)\hat{\mathbf{X}}^{\dagger}(q)\mathbf{X}\mathbf{z}(q) + \hat{\mathbf{X}}(q)\hat{\mathbf{X}}^{\dagger}(q)\mathbf{v}(q).
$$

(3.53)

Taking the Euclidean norm square and expectation on both sides of (3.53), we obtain

$$
\begin{aligned}
E\{\|\hat{\mathbf{X}}(q)\hat{\mathbf{z}}(q+1)\|^2\} &= E\{\|\hat{\mathbf{X}}(q)\hat{\mathbf{z}}(q)\|^2\} - E\{\hat{\mathbf{z}}^H(q)\hat{\mathbf{X}}^H(q)\mathbf{B}\mathbf{X}(q)\mathbf{z}(q)\} \\
&\quad + E\{\|\mathbf{X}(q)\mathbf{z}(q)\|_{\mathbf{A}}^2\} + E\{\|\mathbf{v}(q)\|_{\mathbf{A}}^2\},
\end{aligned}
$$

(3.54)

where $\|\mathbf{v}\|_{\mathbf{A}}^2 = \mathbf{v}^H\mathbf{A}\mathbf{v}$ is the norm of vector $\mathbf{v}$ with weight $\mathbf{A}$ and

$$
\begin{aligned}
\mathbf{A} &= (\hat{\mathbf{X}}^{\dagger})^H\hat{\mathbf{X}}^H\hat{\mathbf{X}}\hat{\mathbf{X}}^{\dagger}, \\
\mathbf{B} &= \hat{\mathbf{X}}\hat{\mathbf{X}}^{\dagger} + (\hat{\mathbf{X}}^{\dagger})^H\hat{\mathbf{X}}^H.
\end{aligned}
$$

(3.55)
(3.56)

Here, the n of $\mathbf{X}(q)$ is omitted for convenience. For steady state, $E\{\|\hat{\mathbf{X}}(q)\hat{\mathbf{z}}(q+1)\|^2\} = E\{\|\hat{\mathbf{X}}(q)\hat{\mathbf{z}}(q)\|^2\}$ and $E\{\hat{\mathbf{z}}^H\hat{\mathbf{X}}^H\mathbf{B}\mathbf{X}\mathbf{z}\} \approx E\{\|\mathbf{X}\mathbf{z}\|_{\mathbf{B}}^2\}$. Therefore

$$
E\{\|\mathbf{X}\mathbf{z}\|_{\mathbf{B}}^2\} - E\{\|\mathbf{X}\mathbf{z}\|_{\mathbf{A}}^2\} \approx E\{\|\mathbf{v}\|_{\mathbf{A}}^2\}.
$$

(3.57)

Using the independence assumption, the right side of (3.57) equals [43]

$$
E\{\|\mathbf{v}\|_{\mathbf{A}}^2\} \approx \sigma_v^2 tr(|E\{\mathbf{A}\}|),
$$

(3.58)

where $tr(.)$ means the trace operation, and the left side of (3.57) equals

$$E\{\|\mathbf{Xz}\|_{\mathbf{B}}^2\} - E\{\|\mathbf{Xz}\|_{\mathbf{A}}^2\} \approx E\{\|\mathbf{Xz}\|^2\}(tr(|E\{\mathbf{B}\}|) - tr(|E\{\mathbf{A}\}|)). \qquad (3.59)$$

Therefore, in steady state

$$E\{\|\mathbf{Xz}\|^2\} \approx \sigma_v^2 \frac{tr(|E\{\mathbf{A}\}|)}{(tr(|E\{\mathbf{B}\}|) - tr(|E\{\mathbf{A}\}|))}. \qquad (3.60)$$

The steady-state MSE of PU LSCMA becomes

$$E\{|e(q)|\} = K\sigma_v^2 + \sigma_v^2 \frac{tr(|E\{\mathbf{A}\}|)}{(tr(|E\{\mathbf{B}\}|) - tr(|E\{\mathbf{A}\}|))}, \quad n \to \infty. \qquad (3.61)$$

Since the MSE of LSCMA is based on an output vector with length K, we can revise this block-based MSE to sample-based MSE as

$$E\{|e(q)|\} = \sigma_v^2 + \sigma_v^2 \frac{tr(|E\{\mathbf{A}\}|)}{K(tr(|E\{\mathbf{B}\}|) - tr(|E\{\mathbf{A}\}|))}, \quad n \to \infty. \qquad (3.62)$$

For the sequential, stochastic, and MMax methods, there is $(tr(|E\{\mathbf{B}\}|) - tr(|E\{\mathbf{A}\}|)) \approx tr(|E\{\mathbf{A}\}|)$. Therefore, the steady-state MSE of the sequential, stochastic, and MMax PU LSCMA are close to the full-update LSCMA.

3.4.3   COMPLEXITY OF THE PU SDCMA AND LSCMA

For an adaptive filter, there are three kinds of computational complexities: filter complexity, adaptive algorithm complexity, and overall complexity for an entire data sequence. In this subsection, we calculate the filter complexity per sample, and adaptive algorithm complexity per sample for PU CMA1-2, NCMA, and LSCMA. The overall complexity is also defined. Since complex-valued data are used, we generally calculate the number of real additions, multiplications (including division), and square-root-divisions (SRD).

The filter complexity means the complexity of (2.8) for SDCMA and (3.9) for LSCMA. Here the complexity of the PU algorithm is the same as the original algorithm. The number of real multiplications in (2.8) is $4N$. The number of real additions in (2.8) is $4N - 2$. The filter complexity of LSCMA is $4NK$ multiplications and $(4N - 2)K$ additions. Note the complexity of (3.9) is for a $K \times 1$ vector. Therefore, the filter complexity per sample is still the same as that of the SDCMA.

The adaptive algorithm complexity means the complexity of the weight update equation. The sequential, stochastic, and MMax PU methods achieve lower complexity than the original algorithms by updating fewer coefficients. The complexity of PU algorithms is obtained from (3.13), (3.15), and (3.16). Note for the periodic partial update method, the complexity is not reduced in this step. Table 3.1 shows the adaptive algorithm complexity per sample of CMA1-2, NCMA, and LSCMA for both PU methods and original algorithms. Note the complexity of different PU methods is not considered here. For fair comparison, the complexity of LSCMA is converted to sample-based complexity, not block complexity. The pseudoinverse of $\hat{\mathbf{X}}^{\dagger}$ in (3.16) can be implemented by eliminating the zero columns and using the QR decomposition of $\hat{\mathbf{X}}$.

Table 3.1: Adaptive algorithm complexity

Algorithms	CMA1-2	NCMA	LSCMA
PU +	4M+3	6M+2	$4M^2 + 2M + \frac{M^2}{K} + 1 - 4\frac{M}{K}$
Org +	4N+3	6N+2	$4N^2 + 2N + \frac{N^2}{K} + 1 - 4\frac{N}{K}$
PU ×	4M+6	6M+7	$4M^2 + 4M + \frac{2M^2}{K} + 4 - \frac{M}{K}$
Org ×	4N+6	6N+7	$4N^2 + 4N + \frac{2N^2}{K} + 4 - \frac{N}{K}$
PU SRD	1	1	$\frac{M}{K} + 1$
Org SRD	1	1	$\frac{N}{K} + 1$

Figures 3.1 and 3.2 show the adaptive algorithm complexity in log-scale as linear combiner dimensionality grows. The full-update length is 64 and the block size K in LSCMA is equal to $4 * 64 = 256$. As the partial update length decreases, the PU LSCMA can reduce the complexity significantly.

The overall complexity means the total computational complexity of an entire data sequence which contains thousands of samples. Table 3.2 shows the overall complexity per sample of NCMA and LSCMA for both PU methods and original algorithms. Fig. 3.3 shows the overall number of digital signal processor (DSP) cycles per sample needed in log-scale as linear combiner dimensionality grows. The full-update length is 64 and the block size K in LSCMA is equal to $4 * 64 = 256$. As the partial-update length decreases, the PU LSCMA can reduce the complexity significantly. We assume each multiplication needs 0.5 cycle, each addition needs 0.5 cycle, and each SDR needs 6 cycles in a TI DSP. The periodic PU method aims to reduce the overall complexity by updating periodically. The LSCMA can stop updating after reaching convergence to reduce the overall computational complexity. Since the convergence rate of LSCMA is much faster than the SDCMA, fewer samples are needed for LSCMA to converge. Therefore, the LSCMA may achieve less overall complexity than the SDCMA.

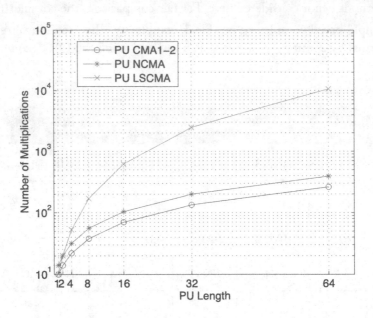

Figure 3.1: The number of multiplications vs. PU length.

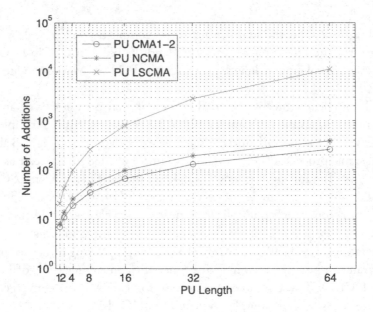

Figure 3.2: The number of additions vs. PU length.

Table 3.2: Overall complexity

Algorithms	CMA1-2	NCMA	LSCMA
PU +	4N+4M+1	4N+6M	$4M^2 + 4N + 2M + \frac{M^2}{K} - 1 - 4\frac{M}{K}$
Org +	8N+1	10N	$4N^2 + 6N + \frac{N^2}{K} - 1 - 4\frac{N}{K}$
PU ×	4N+4M+6	4N+6M+7	$4M^2 + 4N + 4M + \frac{2M^2}{K} + 4 - \frac{M}{K}$
Org ×	8N+6	10N+7	$4N^2 + 8N + \frac{2N^2}{K} + 4 - \frac{N}{K}$
PU SRD	1	1	$\frac{M}{K} + 1$
Org SRD	1	1	$\frac{N}{K} + 1$

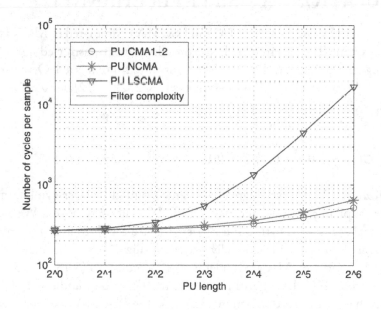

Figure 3.3: The number of cycles vs. PU length.

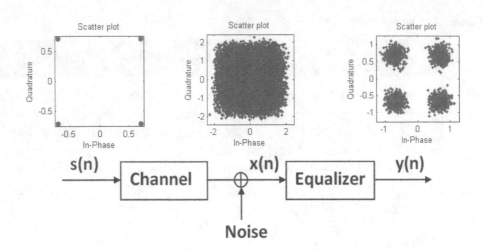

Figure 3.4: Channel equalization model.

3.5 SIMULATION – A SIMPLE FIR CHANNEL

The performance of PU CMA1-2, NCMA, and LSCMA is simulated in Matlab through a channel equalization system. The system model, the constellation of transmitted signal, received signal, and recovered signal are shown in Fig. 3.4. The transmitted signal is a 4-QAM signal. The codewords 00, 01, 10, 11 are modulated to the symbols $(-\sqrt{2}/2, \sqrt{2}/2)$, $(-\sqrt{2}/2, -\sqrt{2}/2)$, $(\sqrt{2}/2, \sqrt{2}/2)$, $(\sqrt{2}/2, -\sqrt{2}/2)$, respectively.

A simple, short FIR channel is used

$$C(z) = -0.3 + 0.4z^{-1} - 0.1z^{-2} + z^{-3} + 0.5z^{-4} - 0.2z^{-5}. \tag{3.63}$$

Without equalization, the received signal has been corrupted by inter-symbol interference (ISI) and noise. After applying an equalizer, most of the ISI will be removed. We assume the entire length of the equalizer is 16 (N=16). The PU lengths use 8 (M=8) and 4 (M=4). The sampling rate is 1, which means it is not a fractionally spaced equalizer. In the equalizer part, different PU CMA1-2, NCMA, and LSCMA are implemented according to (3.13), (3.15), and (3.17). The initial weights of the SDCMA algorithms are set to "1" in the middle and "0" otherwise. The parameters of each algorithm are shown in Table 3.3. The LSCMA does not have step size. The steady-state CMCF, bit-error-rate (BER) performance, and convergence rate are shown.

Table 3.3: The parameters of filter algorithms

Algorithms	Step size (μ)	Initial weights
CMA1-2	0.005	[0,...,1,...0]
NCMA	0.1	[0,...,1,...0]
LSCMA	/	$\neq \mathbf{0}$

3.5.1 CONVERGENCE PERFORMANCE

The CMCF convergence performance is compared among PU CMA1-2, NCMA, and LSCMA. The CMCF results are calculated by using (3.2) and (3.46). The delay and phase offset are not compensated. The periodic CMA has a convergence rate N/M times slower than the original CMA and the performance is not shown here.

Fig. 3.5 shows the convergence performance for both $M = 8$ and $M = 4$. To display the convergence performance clearly, the results are obtained by averaging 1,000 independent runs, and only the first 3,000 samples are shown. The convergence rates vary for different CMA algorithms, different PU methods, and the PU length.

The first row of Fig. 3.5 shows the performance of PU CMA1-2. For the MMax method, the convergence rate is close to that of the original algorithm. For sequential and stochastic methods, the convergence rates are N/M times slower than the original CMA1-2.

The second row shows the performance of PU NCMA. The MMax, sequential, and stochastic PU NCMA have convergence rates close to the NCMA.

The third row shows the performance of PU NCMA. The convergence speed of LSCMA is faster than CMA1-2 and NCMA, as expected. When the PU length is 8, the PU LSCMA algorithms have similar convergence rates. When the PU length is 4, the sequential and stochastic algorithms have convergence rates close to that of the original LSCMA. However, the MMax algorithm has the slowest convergence speed.

The approximate convergence times (number of samples) are listed in Table 3.4 for PU CMA1-2, NCMA, and LSCMA. The PU CMA1-2 has longer convergence times than the original CMA1-2. The MMax NCMA converges a little faster than the original NCMA when PU length is 8. The original LSCMA and the sequential LSCMA can converge faster than the original CMA1-2 and NCMA. The stochastic LSCMA can converge faster than the original CMA1-2 and NCMA when PU length is 4. The MMax LSCMA can converge faster than the original CMA1-2 and NCMA when PU length is 8.

3.5.2 STEADY-STATE PERFORMANCE

The steady state means that the states (i.e., MSE, or BER) of the $(n-1)^{th}$ and n^{th} iterations are similar. There is no significant difference between them. First, we examine the steady-state MSE performance. Table 3.5 shows the theoretical and simulated CMCFs for PU CMA algorithms when SNR is 15 dB. Two cases PU length is equal to 8 ($M = 8$) and PU length is equal

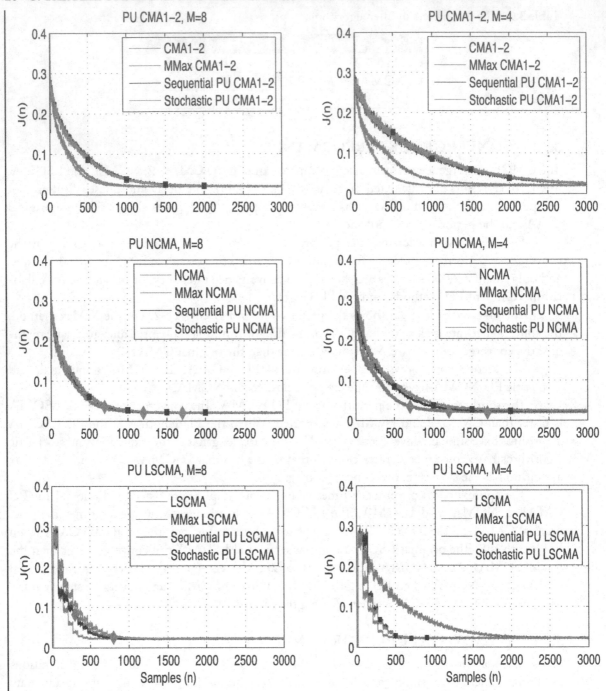

Figure 3.5: The convergence performance among PU CMA algorithms, PU length=4.

Table 3.4: Approximate convergence times

Algorithms	PU length=8	PU length=4
CMA1-2	886	895
MMax CMA1-2	1067	1771
Sequential PU CMA1-2	1763	3307
Stochastic PU CMA1-2	1760	3409
NCMA	1106	1105
MMax NCMA	1052	1230
Sequential PU NCMA	1158	1304
Stochastic PU NCMA	1194	1785
LSCMA	400	402
MMax LSCMA	842	2136
Sequential PU LSCMA	719	588
Stochastic PU LSCMA	981	588

to 4 ($M = 4$), are shown. The theoretical results are calculated according to (3.44), (3.45), and (3.62). The optimal weights are taken from the simulated steady-state weights. Note that the simulated steady-state weights may have a phase rotation to the actually optimal weights. The MSE of LSCMA means the sample-based MSE, not the block-based MSE. The simulated results are calculated by using (3.26) and (3.7). The steady-state MSE is obtained by averaging the last 2,000 samples. We can see that the simulated results match the theoretical results. CMA1-2, NCMA, and LSCMA can achieve similar steady-state MSE. PU CMA algorithms have similar performance to the full-update CMA algorithms. The MSE performance of $M = 4$ is similar to the case of $M = 8$.

The BER performance is also examined for PU CMA algorithms. To calculate BER, the group delay must be compensated. All of the CMA and PU CMA algorithms obtain the same number of delays, which is 10 symbols. A decision-directed algorithm is used to map the symbol to the 4-QAM symbol according to the shortest Euclidean distance rule. Then the 4-QAM symbols are demodulated to the codewords introduced before. The BER is calculated by comparing the codewords between input and recovered signals.

Fig. 3.6 and Fig. 3.7 illustrate the BER in log-scale against SNR between PU CMA1-2 algorithms and the full-length CMA1-2 for $M = 8$ and $M = 4$, respectively. The BER performance of different PU LSCMA algorithms has the same relationship as the CMCF performance, which means the PU CMA1-2 can obtain similar BER as the CMA1-2.

Fig. 3.8 and Fig. 3.9 illustrate the BER performance of PU NCMA for $M = 8$ and $M = 4$, respectively. For $M = 8$, the PU NCMA algorithms can obtain similar BER as the NCMA. However, the PU NCMA perform a little worse than the original NCMA when $M = 4$.

Table 3.5: Steady-state MSE comparison among different PU CMA algorithms

Algorithms	Theory (dB)	Simulation (dB)
PU length=8		
CMA1-2	-14.7034	-15.2515
MMax CMA1-2	-14.8293	-15.2546
Sequential PU CMA1-2	-14.7209	-15.3105
Stochastic PU CMA1-2	-14.7609	-15.3054
PU length=4		
CMA1-2	-14.6937	-15.3041
MMax CMA1-2	-15.0644	-14.6653
Sequential PU CMA1-2	-14.7121	-15.2875
Stochastic PU CMA1-2	-14.7458	-15.2591
PU length=8		
NCMA	-14.5979	-15.2045
MMax NCMA	-14.7053	-15.1209
Sequential PU NCMA	-14.6463	-14.9246
Stochastic PU NCMA	-14.6459	-15.0812
PU length=4		
NCMA	-14.6351	-15.1870
MMax NCMA	-14.7540	-14.9201
Sequential PU NCMA	-14.5777	-14.5603
Stochastic PU NCMA	-14.6601	-14.6230
PU length=8		
LSCMA	-15.289	-15.423
MMax LSCMA	-15.353	-15.525
Sequential PU LSCMA	-15.287	-15.558
Stochastic PU LSCMA	-15.287	-15.525
PU length=4		
LSCMA	-15.5	-15.52
MMax LSCMA	-15.447	-15.696
Sequential PU LSCMA	-15.38	-15.611
Stochastic PU LSCMA	-15.38	-15.669

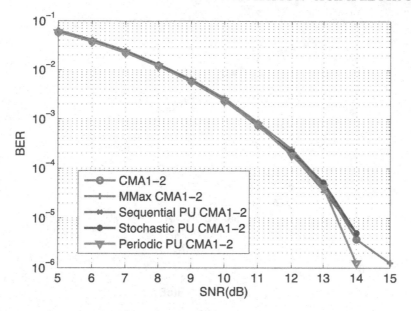

Figure 3.6: Comparison of BER between PU CMA1-2 algorithms and the full-length CMA1-2, PU length=8.

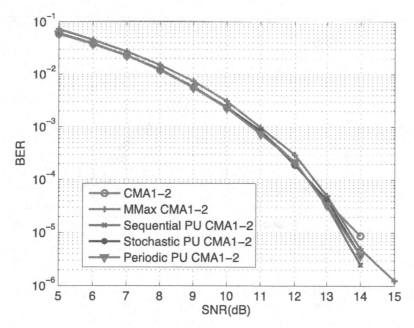

Figure 3.7: Comparison of BER between PU CMA1-2 algorithms and the full-length CMA1-2, PU length=4.

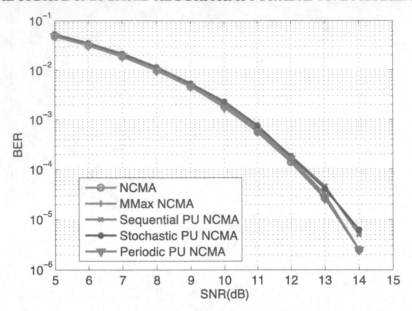

Figure 3.8: Comparison of BER between PU NCMA algorithms and the full-length NCMA, PU length=8.

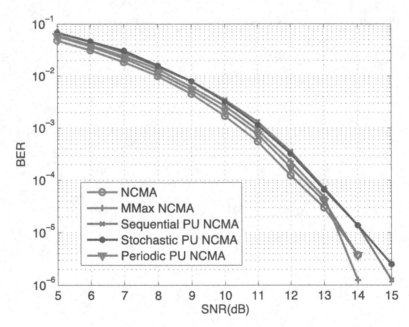

Figure 3.9: Comparison of BER between PU NCMA algorithms and the full-length NCMA, PU length=4.

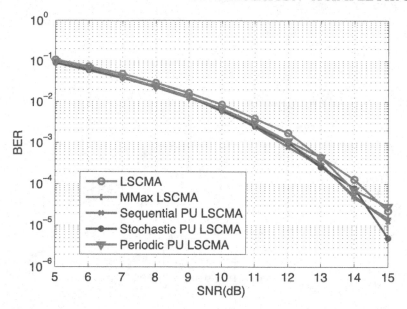

Figure 3.10: Comparison of BER between PU LSCMA algorithms and the full-length LSCMA, PU length=8.

Fig. 3.10 and Fig. 3.11 illustrate the BER performance between PU LSCMA algorithms and the full-length LSCMA for $M = 8$ and $M = 4$, respectively. For $M = 8$, the BER performance of different PU LSCMA algorithms slightly outperforms the full-update LSCMA when SNR is high. For $M = 4$, the PU LSCMA can achieve much better BER performance than the LSCMA at high SNR.

3.5.3 COMPLEXITY

Since partial-update methods can reduce the computational complexity in the adaptive algorithm part, the numerical computational complexity of PU CMA algorithms is compared according to Table 3.1. N is equal to 16. M is equal to 8 and 4, respectively. K in LSCMA is chosen to be 4 times the original equalizer length and is equal to $4 \times 16 = 64$. Table 3.6 shows the adaptive algorithm complexity for the simulation example.

From the table, PU LSCMA can reduce the number of additions or multiplications about $(N/M)^2$ times of the LSCMA, while the PU SC-CMA can only reduce the complexity about N/M times.

From the simulations, we draw the following conclusion:

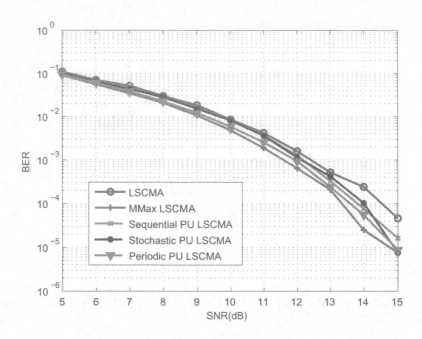

Figure 3.11: Comparison of BER between PU LSCMA algorithms and the full-length LSCMA, PU length=4.

Table 3.6: Adaptive algorithm complexity

Algorithms	CMA1-2	NCMA	LSCMA
Org +	67	98	1060
PUL=8 +	35	50	273.5
PUL=4 +	17	26	73
Org ×	70	103	1099.8
PUL=8 ×	38	55	293.875
PUL=4 ×	22	31	84.4375
Org SRD	1	1	1.25
PUL=8 SRD	1	1	1.125
PUL=4 SRD	1	1	1.0625

1. PU LSCMA can reduce the number of additions or multiplications by about $(N/M)^2$ times that of the LSCMA. PU CMA1-2 and NCMA can reduce the complexity by about N/M times.

2. PU CMA algorithms can achieve similar steady-state performance as the original CMA algorithms.

3. Sequential LSCMA can converge faster than the CMA1-2 and NCMA.

4. PU CMA1-2 and PU NCMA can be used in a real-time processor such as FPGA, and are suitable for a static channel condition.

3.6 ALGORITHM ANALYSIS FOR A TIME-VARYING SYSTEM

In real-world applications, the unknown system is usually time-varying. Therefore, tracking analysis is needed to show how well an adaptive filter performs in a time-varying environment. The optimal weights $\mathbf{w}^o$ become $\mathbf{w}^o(n)$, i.e., they change with time.

A first-order Markov model [43] is used for the time-varying unknown system. It has a form as follows:

$$\mathbf{w}^o(n) = \mathbf{w}^o(n-1) + \mathbf{j}(n), \tag{3.64}$$

where $\mathbf{j}(n)$ is the process noise vector with zero mean and correlation matrix $\mathbf{R_j} = E\{\mathbf{j}^H \mathbf{j}\}$. To determine the tracking performance, another assumption is needed: $\mathbf{j}(n)$ is independent of all other signals.

The coefficient error in the tracking analysis is defined

$$\mathbf{z}(n) = \mathbf{w}^o(n) - \mathbf{w}(n). \tag{3.65}$$

3.6.1 ALGORITHM ANALYSIS OF CMA1-2 AND NCMA FOR A TIME-VARYING SYSTEM

Use the same analyses in time-invariant Section 3.4.1. The coefficient error of CMA1-2 in (3.33) becomes

$$\hat{\mathbf{z}}(n+1) = \hat{\mathbf{z}}(n) + \mathbf{j}(n+1) + \mu\hat{\mathbf{x}}^*(n)v(n) - \mu\hat{\mathbf{x}}^*(n)\mathbf{x}^T(n)\mathbf{z}(n). \tag{3.66}$$

Take the expectation of (3.66). Since $E\{v\} = 0$, $E\{\mathbf{z}\} = \mathbf{0}$, and $E\{\mathbf{j}\} = \mathbf{0}$, the average weights still converge to the optimal weights at steady state. The convergence condition is

$$0 < \mu < \frac{2}{\beta\lambda_{max}(E\{\mathbf{x}^*\mathbf{x}^T\})}, \tag{3.67}$$

$$\begin{cases} \beta \approx \frac{M}{N}, & \text{for sequential and stochastic methods} \\ \frac{M}{N} < \beta < 1, & \text{for MMax method} \end{cases} \qquad (3.68)$$

Following the analyses from (3.39) to (3.42), the steady-state MSE of PU CMA1-2 in a time-varying system becomes

$$E\{|e(n)|^2\} = \sigma_v^2 + \frac{\mu \sigma_v^2 E\{\|\hat{\mathbf{x}}\|^2\} + \mu^{-1} E\{\|\mathbf{J}\|^2\}}{2\beta - \mu E\{\|\hat{\mathbf{x}}\|^2\}}, \qquad n \to \infty. \qquad (3.69)$$

Assuming the process noise is white with variance σ_η^2, the steady-state MSE of PU CMA1-2 in a time-varying system can be simplified as

$$E\{|e(n)|^2\} = \sigma_v^2 + \frac{\mu \sigma_v^2 E\{\|\hat{\mathbf{x}}\|^2\} + \mu^{-1} N \sigma_\eta^2}{2\beta - \mu E\{\|\hat{\mathbf{x}}\|^2\}}, \qquad n \to \infty. \qquad (3.70)$$

For PU NCMA, the convergence in the mean is achieved if $0 < \mu < 2$. The steady-state MSE of PU NCMA in a time-varying system becomes

$$E\{|e(n)|^2\} = \sigma_v^2 + \frac{\mu \sigma_v^2 + \mu^{-1} tr(\mathbf{R_J} E\{\hat{\mathbf{x}}^H \hat{\mathbf{x}}\})}{2 - \mu}, \qquad n \to \infty. \qquad (3.71)$$

If white process noise is used, it can be simplified as

$$E\{|e(n)|^2\} = \sigma_v^2 + \frac{\mu \sigma_v^2 + \mu^{-1} \sigma_\eta^2 E\{\|\hat{\mathbf{x}}\|^2\}}{2 - \mu}, \qquad n \to \infty. \qquad (3.72)$$

3.6.2   ALGORITHM ANALYSIS OF LSCMA FOR A TIME-VARYING SYSTEM

Use the same analyses in time-invariant Section 3.4.2. The coefficient error of LSCMA in (3.50) becomes

$$\hat{\mathbf{z}}(q + 1) = \hat{\mathbf{z}}(q) + \mathbf{J}(q + 1) + \hat{\mathbf{X}}^\dagger(q)\mathbf{v}(q) - \hat{\mathbf{X}}^\dagger(q)\mathbf{X}(q)\mathbf{z}(q). \qquad (3.73)$$

Take the expectation of (3.73). Since $E\{v\} = 0$, $E\{\mathbf{z}\} = \mathbf{0}$, and $E\{\mathbf{J}\} = \mathbf{0}$, the average weights still converge to the optimal weights at steady state. Convergence in the mean is achieved if the inversion of $E\{\hat{\mathbf{X}}^\dagger \mathbf{X}\}$ exists, which means $E\{\hat{\mathbf{X}}^\dagger \mathbf{X}\}$ has full rank and it is nonsingular. Again, follow the analyses from (3.53) to (3.60), the steady-state MSE of PU LSCMA in a time-varying system becomes

$$E\{|e(q)|^2\} = K\sigma_v^2 + \frac{\sigma_v^2 tr(|E\{\mathbf{A}\}|) + tr(\mathbf{R_J} E\{\hat{\mathbf{X}}^H \hat{\mathbf{X}}\})}{(tr(|E\{\mathbf{B}\}|) - tr(|E\{\mathbf{A}\}|))}, \qquad n \to \infty. \qquad (3.74)$$

Since the cost function of LSCMA is based on an output vector with length K, the block-based MSE can be revised to the sample-based MSE as

$$E\{|e(q)|^2\} = \sigma_v^2 + \frac{\sigma_v^2 tr(|E\{\mathbf{A}\}|) + tr(\mathbf{R_J} E\{\hat{\mathbf{X}}^H \hat{\mathbf{X}}\})}{K(tr(|E\{\mathbf{B}\}|) - tr(|E\{\mathbf{A}\}|))}, \qquad n \to \infty. \qquad (3.75)$$

If the white process noise is used, it can be simplified as

$$E\{|e(q)|^2\} = \sigma_v^2 + \frac{\sigma_v^2 tr(|E\{\mathbf{A}\}|) + \sigma_\eta^2 tr(E\{\hat{\mathbf{X}}^H \hat{\mathbf{X}}\})}{K(tr(|E\{\mathbf{B}\}|) - tr(|E\{\mathbf{A}\}|))}, \quad n \to \infty. \tag{3.76}$$

3.6.3 SIMULATION

The same simulation example as in Section 3.5 is used, except the FIR channel changes with time. The process noise is white noise with $\sigma_\eta = 0.001$ and $\sigma_\eta = 0.0001$. SNR is 15 dB. The results are obtained by averaging 100 independent runs.

Fig. 3.12 shows the CMCF performance for PU CMA1-2 with PU length $M = 8$ and $\sigma_\eta = 0.001$. The MMax, sequential, and stochastic CMA1-2 perform similar to the full-update CMA1-2 at steady state. Fig. 3.13 shows the CMCF performance for PU CMA1-2 with PU length $M = 4$ and $\sigma_\eta = 0.001$. In this case, the MMax, sequential, and stochastic CMA1-2 perform worse than the full-update CMA1-2. This is because the process noise dominates the MSE and it increases as the PU length reduces according to the theoretical results. Step size μ is 0.0025 for $\sigma_\eta = 0.001$. Fig. 3.14 shows the CMCF performance for PU CMA1-2 with PU length $M = 8$ and $\sigma_\eta = 0.0001$. All the PU CMA1-2 and the original CMA1-2 have similar steady-state performance. Fig. 3.15 shows the CMCF performance for PU CMA1-2 with PU length $M = 4$ and $\sigma_\eta = 0.0001$. Again, all the PU CMA1-2 and the original CMA1-2 have similar steady-state performance. Comparing Fig. 3.14 and Fig. 3.15 with Fig. 3.12 and Fig. 3.13, we can see that the CMCF increases as the process noise increases. Step size μ for $\sigma_\eta = 0.0001$ is 0.005.

Fig. 3.16 shows the CMCF performance for PU NCMA with PU length $M = 8$ and $\sigma_\eta = 0.001$. Fig. 3.17 shows the CMCF performance for PU NCMA with PU length $M = 4$ and $\sigma_\eta = 0.001$. The sequential and stochastic NCMA have the lower CMCF than the full-update NCMA in these two cases. This is because the steady-state MSE of PU NCMA is related to $E\{\|\hat{\mathbf{x}}\|^2\}$. The average norm-square of the sequential and stochastic PU input vector is smaller than the average norm-square of the full-update input vector. When process noise is big, the steady-state MSE of the sequential and stochastic NCMA is smaller than the full-update NCMA. The step size μ in the case $\sigma_\eta = 0.001$ is 0.01. Fig. 3.18 shows the CMCF performance for PU NCMA with PU length $M = 8$ and $\sigma_\eta = 0.0001$. Fig. 3.19 shows the CMCF performance for PU NCMA with PU length $M = 4$ and $\sigma_\eta = 0.0001$. The step size μ in the case $\sigma_\eta = 0.0001$ is 0.01. Since the process noise is very small, the performance of PU length 4 is similar to that of PU length 8, and the performance of PU NCMA is similar to full-update NCMA at steady state. Decreasing the PU length does not necessarily decrease the performance.

Fig. 3.20 shows the CMCF performance for PU LSCMA with PU length $M = 8$ and $\sigma_\eta = 0.001$. In this case, the PU LSCMA has similar performance to the full-update LSCMA. Fig. 3.21 shows the CMCF performance for PU LSCMA with PU length $M = 4$ and $\sigma_\eta = 0.001$. The PU LSCMA still has similar steady-state MSE as the full-update LSCMA. Fig. 3.22

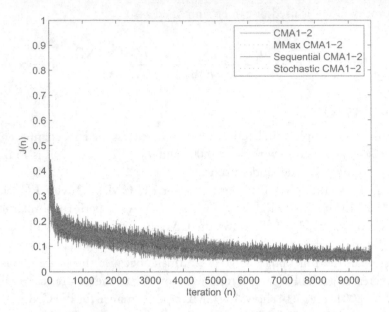

Figure 3.12: The CMCF performance for PU CMA1-2 for a time-varying system, PU length=8, $\sigma_\eta = 0.001$.

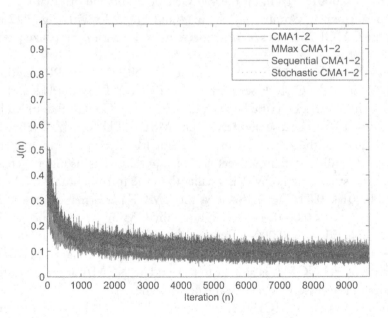

Figure 3.13: The CMCF performance for PU CMA1-2 for a time-varying system, PU length=4, $\sigma_\eta = 0.001$.

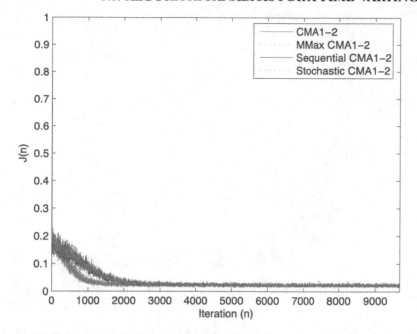

Figure 3.14: The CMCF performance for PU CMA1-2 for a time-varying system, PU length=8, $\sigma_\eta = 0.0001$.

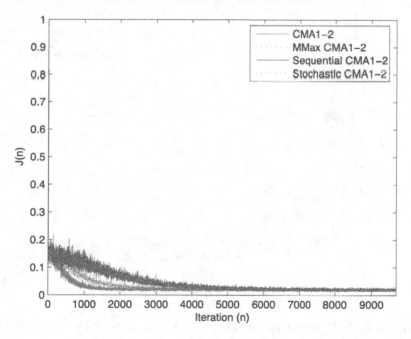

Figure 3.15: The CMCF performance for PU CMA1-2 for a time-varying system, PU length=4, $\sigma_\eta = 0.0001$.

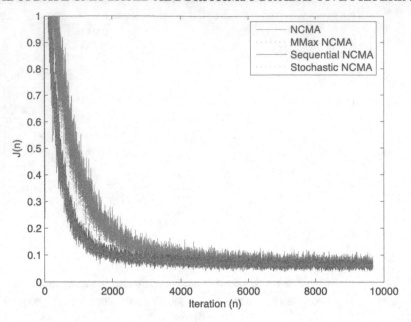

Figure 3.16: The CMCF performance for PU NCMA for a time-varying system, PU length=8, $\sigma_\eta = 0.001$.

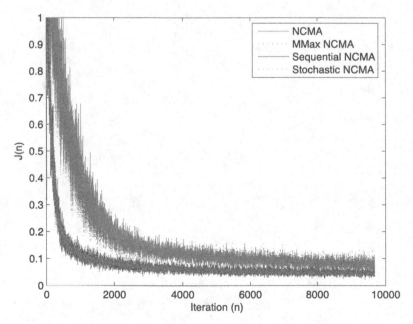

Figure 3.17: The CMCF performance for PU NCMA for a time-varying system, PU length=4, $\sigma_\eta = 0.001$.

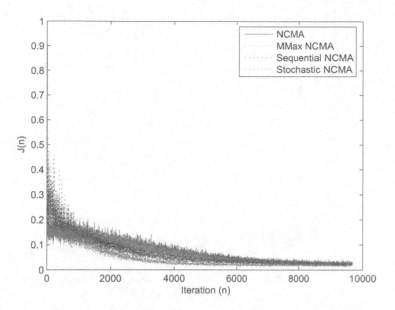

Figure 3.18: The CMCF performance for PU NCMA for a time-varying system, PU length=8, $\sigma_\eta = 0.0001$.

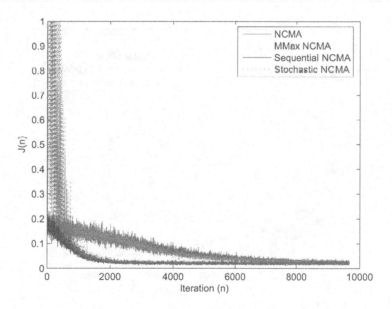

Figure 3.19: The CMCF performance for PU NCMA for a time-varying system, PU length=4, $\sigma_\eta = 0.0001$.

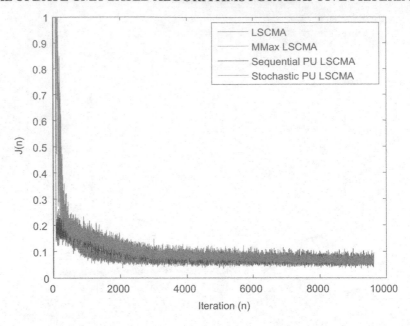

Figure 3.20: The CMCF performance for PU dynamic LSCMA for a time-varying system, PU length=8, $\sigma_\eta = 0.001$.

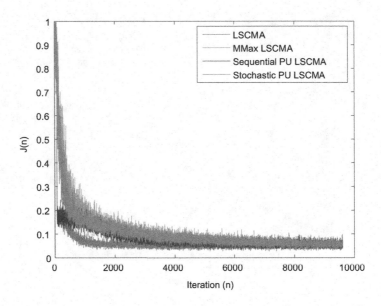

Figure 3.21: The CMCF performance for PU dynamic LSCMA for a time-varying system, PU length=4, $\sigma_\eta = 0.001$.

shows the CMCF performance for PU LSCMA with PU length $M = 8$ and $\sigma_\eta = 0.0001$. Fig. 3.23 shows the CMCF performance for PU LSCMA with PU length $M = 4$ and $\sigma_\eta = 0.0001$. Since the process noise is small, the PU LSCMA has similar steady-state performance to the full-update LSCMA. We can see that the CMCF increases as the process noise increases. Since the PU LSCMA can achieve similar tracking performance as the full-update LSCMA while reducing the computational complexity significantly, it is suitable for systems with dynamic environments.

The steady-state results at PU length is equal to 4 are more fuzzy than those at PU length is equal to 8 for all the PU CMA-based algorithms. This is because tracking performance is a non-stationary system and the temporal variance of MSE increases as the degree of non-stationary increases [22].

Tables 3.7 and 3.8 show the theoretical and simulated MSE for PU CMA algorithms when $\sigma_\eta = 0.001$ and $\sigma_\eta = 0.0001$, respectively. Two cases, PU length is equal to 8 ($M = 8$) and PU length is equal to 4 ($M = 4$), are shown. The theoretical results are calculated by (3.70), (3.72), and (3.76). We can see that the simulated results match the theoretical results.

3.7   CONCLUSION

In this chapter, PU methods are applied to the CMA-based algorithms including CMA1-2, NCMA, and LSCMA. Theoretical mean and mean-square performance are derived for both time-invariant and time-varying systems. The performance of different PU CMA-based algorithms are compared by using computer simulations. The simulated results match the theoretical results. The PU CMA-based algorithms can achieve steady-state MSE similar to the full-update CMA-based algorithms in a time-invariant system and a simple FIR channel. Among different PU methods, the MMax method usually has a convergence rate very close to the full-update method. The sequential and stochastic methods usually converge more slowly than the MMax method. However, the MMax method does not always perform well with the LSCMA algorithm. When the PU length is short, the MMax LSCMA has a slow convergence rate. The sequential LSCMA usually has the best performance among the PU LSCMA algorithms. The PU NCMA algorithms may perform better than the full-update NCMA algorithm in tracking a time-varying system. The PU LSCMA can achieve similar tracking performance as the full-update LSCMA while reducing the computational complexity significantly; it is suitable for systems with dynamic environments.

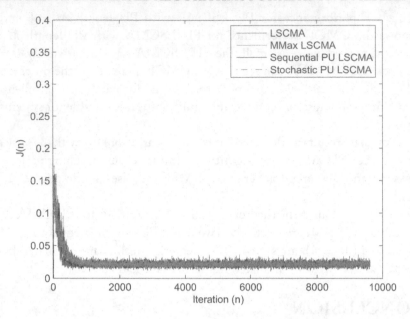

Figure 3.22: The CMCF performance for PU dynamic LSCMA for a time-varying system, PU length=8, $\sigma_\eta = 0.0001$.

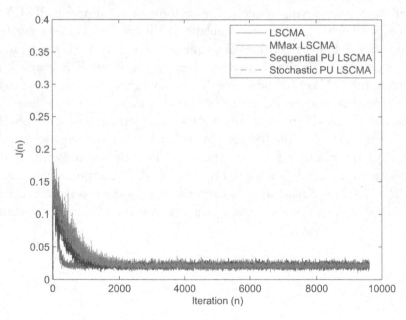

Figure 3.23: The CMCF performance for PU dynamic LSCMA for a time-varying system, PU length=4, $\sigma_\eta = 0.0001$.

Table 3.7: Steady-state MSE comparison among different PU CMA algorithms for a time-varying system, $\sigma_\eta = 0.001$

Algorithms	Theory (dB)	Simulation (dB)
PU length=8		
CMA1-2	-12.997	-12.345
MMax CMA1-2	-12.115	-11.703
Sequential PU CMA1-2	-11.492	-11.506
Stochastic PU CMA1-2	-11.492	-11.591
PU length=4		
CMA1-2	-11.885	-11.563
MMax CMA1-2	-9.419	-9.649
Sequential PU CMA1-2	-9.828	-9.9328
Stochastic PU CMA1-2	-9.835	-9.8644
PU length=8		
NCMA	-13.421	-12.843
MMax NCMA	-13.61	-12.933
Sequential PU NCMA	-13.783	-13.158
Stochastic PU NCMA	-13.779	-13.138
PU length=4		
NCMA	-11.217	-11.034
MMax NCMA	-12.194	-11.548
Sequential PU NCMA	-13.081	12.9
Stochastic PU NCMA	-13.091	-12.86
PU length=8		
LSCMA	-11.96	-11.858
MMax LSCMA	-12.046	-11.779
Sequential PU LSCMA	-12.007	-11.607
Stochastic PU LSCMA	-11.856	-11.709
PU length=4		
LSMCA	-12.239	-12.239
MMax LSCMA	-12.586	-12.181
Sequential PU LSCMA	-12.619	-12.554
Stochastic PU LSCMA	-12.625	-12.391

Table 3.8: Steady-state MSE comparison among different PU CMA algorithms for a time-varying system, $\sigma_\eta = 0.0001$

Algorithms	Theory (dB)	Simulation (dB)
PU length=8		
CMA1-2	-16.159	-16.464
MMax CMA1-2	-16.239	-16.416
Sequential PU CMA1-2	-16.295	-16.482
Stochastic PU CMA1-2	-16.298	-16.476
PU length=4		
CMA1-2	-16.388	-17.073
MMax CMA1-2	-16.614	-16.896
Sequential PU CMA1-2	-16.583	-16.926
Stochastic PU CMA1-2	-16.574	-16.936
PU length=8		
NCMA	-13.824	-13.978
MMax NCMA	-14.075	-14.283
Sequential PU NCMA	-14.609	-14.966
Stochastic PU NCMA	-14.558	-15.116
PU length=4		
NCMA	-13.871	-13.843
MMax NCMA	-14.212	-13.827
Sequential PU NCMA	-14.623	-14.72
Stochastic PU NCMA	-14.625	-14.653
PU length=8		
LSCMA	-16.698	-16.663
MMax LSCMA	-16.789	-16.754
Sequential PU LSCMA	-16.702	-16.735
Stochastic PU LSCMA	-16.708	-16.758
PU length=4		
LSCMA	-15.014	-15.535
MMax LSCMA	-15.15	-15.694
Sequential PU LSCMA	-15.027	-15.691
Stochastic PU LSCMA	-15.022	-15.723

CHAPTER 4

Partial-Update CG Algorithms for Adaptive Filtering

In this chapter, partial-update techniques are applied to the CG algorithm. CG solves the same cost function as the RLS algorithm. It has a fast convergence rate and can achieve the same mean-square performance as RLS at steady state. It has lower computational complexity when compared with the RLS algorithm. The basic partial update methods such as sequential PU, stochastic PU, and MMax update method, are applied to the CG algorithm. There is another PU method, periodic method, which updates the adaptive filter periodically. Since the steady-state performance of the periodic method is the same as the full-update method, it is not considered here. The mean and mean-square performance of different PU CG for a time-invariant system are analyzed and compared with the full-update CG algorithm. The goal of this chapter is to find one or more PU CG algorithms which can reduce the computational complexity while maintaining good performance. The tracking performance of the PU CG for a time-varying system is also studied. Theoretical MSE results are derived for the PU CG at steady state. Computer simulation results are also presented to show the tracking performance of the MMax CG. The performance of the MMax CG is also compared with the full-update CG, full-update RLS, and MMax RLS. The analysis of time-varying systems is necessary because the unknown systems in system identification, echo cancellation, and channel equalization are often time-varying in real world applications.

4.1   REVIEW OF CONJUGATE GRADIENT ALGORITHM

In this chapter, only the CG with the non-reset Polak-Ribière (PR) method is considered since the CG algorithm with the reset method needs higher computational complexity than the non-reset method. The CG with the PR method [7] is summarized as follows:

Initial conditions: $\mathbf{w}(0) = \mathbf{0}$, $\mathbf{R}(n) = \mathbf{0}$, $\mathbf{p}(1) = \mathbf{g}(0)$.

$$e(n) = d(n) - \mathbf{x}^T(n)\mathbf{w}(n-1), \tag{4.1}$$

$$\mathbf{R}(n) = \lambda\mathbf{R}(n-1) + \mathbf{x}(n)\mathbf{x}^T(n), \tag{4.2}$$

$$\alpha(n) = \varphi\frac{\mathbf{p}^T(n)\mathbf{g}(n-1)}{\mathbf{p}^T(n)\mathbf{R}(n)\mathbf{p}(n)}, \tag{4.3}$$

$$\mathbf{w}(n) = \mathbf{w}(n-1) + \alpha(n)\mathbf{p}(n), \tag{4.4}$$

$$\mathbf{g}(n) = \lambda\mathbf{g}(n-1) - \alpha(n)\mathbf{R}(n)\mathbf{p}(n) + \mathbf{x}(n)e(n), \tag{4.5}$$

$$\beta(n) = \frac{(\mathbf{g}(n) - \mathbf{g}(n-1))^T \mathbf{g}(n)}{\mathbf{g}^T(n-1)\mathbf{g}(n-1)}, \tag{4.6}$$

$$\mathbf{p}(n+1) = \mathbf{g}(n) + \beta(n)\mathbf{p}(n), \tag{4.7}$$

where $\mathbf{R}$ is the time-average correlation matrix of $\mathbf{x}$, $\mathbf{p}$ is the search direction, and $\mathbf{g}$ is the residue vector which is also equal to $\mathbf{b}(n) - \mathbf{R}(n)\mathbf{w}(n)$, where $\mathbf{b}(n) = \lambda\mathbf{b}(n-1) + \mathbf{x}(n)d(n)$ is the estimated crosscorrelation of $\mathbf{x}$ and d. The choice of $\mathbf{g}(0)$ can be $d(1)\mathbf{x}(1)$ or satisfies $\mathbf{g}^T(0)\mathbf{g}(0) = 1$. λ is the forgetting factor and the constant parameter η satisfies $\lambda - 0.5 \leq \eta \leq \lambda$.

4.2   PARTIAL-UPDATE CG

The partial-update CG algorithm is summarized as follows:

$$e(n) = d(n) - \mathbf{x}^T(n)\mathbf{w}(n-1), \tag{4.8}$$

$$\hat{\mathbf{R}}(n) = \lambda\hat{\mathbf{R}}(n-1) + \hat{\mathbf{x}}(n)\hat{\mathbf{x}}^T(n), \tag{4.9}$$

$$\alpha(n) = \varphi\frac{\mathbf{p}^T(n)\mathbf{g}(n-1)}{\mathbf{p}^T(n)\hat{\mathbf{R}}(n)\mathbf{p}(n)}, \tag{4.10}$$

$$\mathbf{w}(n) = \mathbf{w}(n-1) + \alpha(n)\mathbf{p}(n), \tag{4.11}$$

$$\mathbf{g}(n) = \lambda\mathbf{g}(n-1) - \alpha(n)\hat{\mathbf{R}}(n)\mathbf{p}(n) + \hat{\mathbf{x}}(n)e(n), \tag{4.12}$$

$$\beta(n) = \frac{(\mathbf{g}(n) - \mathbf{g}(n-1))^T \mathbf{g}(n)}{\mathbf{g}^T(n-1)\mathbf{g}(n-1)}, \tag{4.13}$$

$$\mathbf{p}(n+1) = \mathbf{g}(n) + \beta(n)\mathbf{p}(n), \tag{4.14}$$

where

$$\hat{\mathbf{x}} = \mathbf{I}_M\mathbf{x}, \tag{4.15}$$

and

$$\mathbf{I}_M(n) = \begin{bmatrix} i_1(n) & 0 & \dots & 0 \\ 0 & i_2(n) & \ddots & \vdots \\ \vdots & \ddots & \ddots & 0 \\ 0 & \dots & 0 & i_N(n) \end{bmatrix}, \tag{4.16}$$

$$\sum_{k=1}^{N} i_k(n) = M, \quad i_k(n) \in \{0, 1\}, \tag{4.17}$$

For each iteration, only M elements of the input vector are used to update the weights. Note, the calculation of the output signal error still uses the entire input vector, not the subselected input vector.

The computational complexity of different PU CG for real signals is shown in Table 4.1. The MMax CG uses the SORTLINE sorting method and introduces $2 \lceil log_2 N \rceil + 2$ comparisons [11]. The sequential method does not need extra additions or multiplications. The stochastic CG uses the linear congruential generator and introduces two extra multiplications and two extra additions [11]. Table 4.2 shows the computational complexity of different PU CG for complex signals.

Table 4.1: The computational complexities of PU CG for real signals

Algorithms	Number of multiplications per sample	Number of additions per sample	Comparisons
CG	$3N^2 + 10N + 3$	$2N^2 + 8N - 3$	0
Sequential CG	$2N^2 + M^2 + 9N + M + 3$	$N^2 + M^2 + 8N - 3$	0
Stochastic CG	$2N^2 + M^2 + 9N + M + 5$	$N^2 + M^2 + 8N - 1$	0
MMax CG	$2N^2 + M^2 + 9N + M + 3$	$N^2 + M^2 + 8N - 3$	$2 \lceil log_2 N \rceil + 2$

4.3   STEADY-STATE PERFORMANCE OF PARTIAL-UPDATE CG FOR A TIME-INVARIANT SYSTEM

The normal equation of the partial-update CG algorithm can be represented as:

$$\mathbf{X}_M^T(n) \Lambda(n) \mathbf{X}(n) \mathbf{w}(n) = \mathbf{X}_M^T(n) \Lambda(n) \mathbf{d}(n), \tag{4.18}$$

where $\mathbf{d}(n) = [d(n), d(n-1), \ldots, d(1)]^T$,

$$\mathbf{X}_M(n) = \begin{bmatrix} \hat{\mathbf{x}}_M^T(n) \\ \hat{\mathbf{x}}_M^T(n-1) \\ \vdots \\ \hat{\mathbf{x}}_M^T(1) \end{bmatrix}, \tag{4.19}$$

Table 4.2: The computational complexities of PU CG for complex signals

Algorithms	Number of multiplications per sample	Number of additions per sample	Comparisons
CG	$10N^2 + 38N + 10$	$10N^2 + 43N$	0
Sequential CG	$6N^2 + 4M^2 + 34N + 4M + 10$	$5N^2 + 5M^2 + 40N + 3M$	0
Stochastic CG	$6N^2 + 4M^2 + 34N + 4M + 12$	$5N^2 + 5M^2 + 40N + 3M + 2$	0
MMax CG	$6N^2 + 4M^2 + 36N + 4M + 10$	$5N^2 + 5M^2 + 41N + 3M$	$2 \lceil log_2 \rceil + 2$

and

$$\Lambda(n) = \begin{bmatrix} 1 & 0 & \cdots & 0 \\ 0 & \lambda & \ddots & \vdots \\ \vdots & \ddots & \ddots & 0 \\ 0 & \cdots & 0 & \lambda^n \end{bmatrix}. \tag{4.20}$$

Therefore, the residue vector $\mathbf{g}$ can also be written as

$$\mathbf{g} = \hat{\mathbf{b}}(n) - \widetilde{\mathbf{R}}(n)\mathbf{w}(n), \tag{4.21}$$

where

$$\begin{aligned} \widetilde{\mathbf{R}}(n) &= \lambda \widetilde{\mathbf{R}}(n-1) + \hat{\mathbf{x}}(n)\mathbf{x}^T(n), & (4.22) \\ \hat{\mathbf{b}}(n) &= \lambda \hat{\mathbf{b}}(n-1) + \hat{\mathbf{x}}(n)d(n). & (4.23) \end{aligned}$$

To simplify the analysis, we assume that the input signal is wide-sense stationary and ergodic, and $\alpha(n)$, $\beta(n)$, $\widetilde{\mathbf{R}}(n)$, and $\mathbf{w}(n)$ are uncorrelated to each other [7]. Apply the expectation operator to (4.11), (4.12), and (4.14). Define $E\{\alpha(n)\} = \bar{\alpha}$, $E\{\beta(n)\} = \bar{\beta}$, $E\{\hat{\mathbf{b}}(n)\} = \hat{\mathbf{b}}$, and $E\{\widetilde{\mathbf{R}}(n)\} = \widetilde{\mathbf{R}}$. The system can be viewed as linear and time invariant at steady state. Therefore, the $\mathcal{Z}$-transform can be applied to the system. Define $\mathbf{W}(z) = \mathcal{Z}\{E\{\mathbf{w}(n)\}\}$, $\mathbf{G}(z) = \mathcal{Z}\{E\{\mathbf{g}(n)\}\}$, and $\mathbf{P}(z) = \mathcal{Z}\{E\{\mathbf{p}(n)\}\}$. Equations (4.11), (4.12), and (4.14) become

$$\begin{aligned} \mathbf{W}(z) &= \mathbf{W}(z)z^{-1} + \bar{\alpha}\mathbf{P}(z), & (4.24) \\ \mathbf{G}(z) &= \lambda \mathbf{G}(z)z^{-1} - \bar{\alpha}\hat{\mathbf{R}}\mathbf{P}(z) - \widetilde{\mathbf{R}}\mathbf{W}(z)z^{-1} + \frac{\hat{\mathbf{b}}z}{z-1}, & (4.25) \\ z\mathbf{P}(z) &= \mathbf{G}(z) + \bar{\beta}\mathbf{P}(z). & (4.26) \end{aligned}$$

Therefore,

$$\mathbf{W}(z) = \{[(z-1)(z-\beta)\mathbf{I} + \alpha\hat{\mathbf{R}}z](z-1) + \alpha\widetilde{\mathbf{R}}z\}^{-1}\frac{\alpha\hat{\mathbf{b}}z^3}{z-1}. \tag{4.27}$$

Since the system is causal and $n \geq 0$, the $\mathcal{Z}$-transform is one-sided and $\mathbf{W}(z) = \mathbf{W}^+(z)$. At steady state, the mean of the weights converges to

$$\lim_{n\to\infty} E\{\mathbf{w}(n)\} = \lim_{z\to 1} \mathbf{W}^+(z) = \widetilde{\mathbf{R}}^{-1}\hat{\mathbf{b}}. \tag{4.28}$$

For the causal system to be stable, all the poles must be inside the unit circle. Therefore, the conditions for the stability are $|\bar{\beta}| < 1$, $\widetilde{\lambda}_{min} > 0$, and $0 \leq \bar{\alpha} \leq \frac{2(1+\lambda)(\bar{\beta}+1)}{2\lambda_{max}-\lambda_{min}}$, where $\widetilde{\lambda}_{max}$ is the maximal eigenvalue of $\widetilde{\mathbf{R}}$ and $\widetilde{\lambda}_{min}$ is the minimal eigenvalue of $\widetilde{\mathbf{R}}$. For the sequential and stochastic methods, the partial update correlation matrix $\hat{\mathbf{R}}$ may become ill-conditioned and the condition

$\widetilde{\lambda}_{min} > 0$ cannot be satisfied, especially when M becomes smaller, and the algorithm may suffer convergence difficulty.

Since the input noise $v(n)$ is assumed to be zero mean white noise and independent of the input signal $\mathbf{x}(n)$, the MSE equation of the PU CG algorithm becomes

$$E\{|e(n)|^2\} = \sigma_v^2 + tr(\mathbf{R}E\{(n)^T(n)\}), \tag{4.29}$$

where $\sigma_v^2 = E\{v^2(n)\}$ is the variance of the input noise, and $(n) = \mathbf{w}^* - \mathbf{w}(n)$ is the weight error vector. To simplify the analysis, it is also assumed that the weight error (n) is independent of the input signal $\mathbf{x}(n)$ at steady state.

At steady state,

$$\mathbf{w}(n) \approx \widetilde{\mathbf{R}}^{-1}(n)\hat{\mathbf{b}}(n). \tag{4.30}$$

Using (4.22), (4.23), and (2.4), $\mathbf{w}(n)$ can be written as

$$\mathbf{w}(n) \approx \mathbf{w}^* + \widetilde{\mathbf{R}}^{-1}(n) \sum_{i=1}^{n} \lambda^{n-i} \hat{\mathbf{x}}(i) v(i). \tag{4.31}$$

Define the weight error correlation matrix as

$$\begin{aligned} \mathbf{K}(n) &= E\{(n)^T(n)\} \\ &= E\{(\mathbf{w}^* - \mathbf{w}(n))(\mathbf{w}^* - \mathbf{w}(n))^T\}. \end{aligned} \tag{4.32}$$

Substituting (4.31) into (4.32) and applying the assumptions, we get

$$\mathbf{K}(n) \approx E\{\widetilde{\mathbf{R}}^{-1}(n) \sum_{i=1}^{n} \sum_{j=1}^{n} \lambda^{n-i} \lambda^{n-j} \hat{\mathbf{x}}(i) \hat{\mathbf{x}}^T(j) \widetilde{\mathbf{R}}^{-T}(n)\} E\{v(i)v(j)\}. \tag{4.33}$$

Since the input noise is white,

$$E\{v(i)v(j)\} = \begin{cases} \sigma_v^2 & \text{for } i = j \\ 0 & \text{otherwise} \end{cases}. \tag{4.34}$$

Therefore, $\mathbf{K}(n)$ becomes

$$\mathbf{K}(n) \approx \sigma_v^2 E\{\widetilde{\mathbf{R}}^{-1}(n) \sum_{i=1}^{n} \lambda^{2(n-i)} \hat{\mathbf{x}}(i) \hat{\mathbf{x}}^T(i) \widetilde{\mathbf{R}}^{-T}(n)\} \tag{4.35}$$

for the correlated input signal, and it becomes

$$\mathbf{K}(n) \approx \sigma_v^2 \widetilde{\mathbf{R}}^{-1} \widehat{\widetilde{\mathbf{R}}} \widetilde{\mathbf{R}}^{-T}, \tag{4.36}$$

for the white input signal, where $\widehat{\widehat{\mathbf{R}}} = E\{\sum_{i=1}^{n} \lambda^{2(n-i)} \hat{\mathbf{x}}(i) \hat{\mathbf{x}}^T(i)\}$.

The MSE equation becomes

$$E\{|e(n)|^2\} \approx \sigma_v^2 + \sigma_v^2 tr(\mathbf{R} E\{\widetilde{\mathbf{R}}^{-1}(n) \sum_{i=1}^{n} \lambda^{2(n-i)} \hat{\mathbf{x}}(i) \hat{\mathbf{x}}^T(i) \widetilde{\mathbf{R}}^{-T}(n)\}), \tag{4.37}$$

for the correlated input signal, and it becomes

$$E\{|e(n)|^2\} \approx \sigma_v^2 + \frac{N(1-\lambda)}{1+\lambda} \sigma_v^2 \sigma_x^2 \sigma_{\hat{x}}^2 \sigma_{\widetilde{x}}^{-4} \tag{4.38}$$

for the white input signal, where $tr(\cdot)$ is the trace operator, $\sigma_x^2 = tr(\mathbf{R})$ is the variance of the white input signal, $\sigma_{\hat{x}}^2 = tr(\hat{\mathbf{R}})$, and $\sigma_{\widetilde{x}}^2 = tr(\widetilde{\mathbf{R}})$.

4.4   STEADY-STATE PERFORMANCE OF PARTIAL-UPDATE CG FOR A TIME-VARYING SYSTEM

In a non-stationary environment, the unknown system is time-varying. The desired signal can be rewritten as

$$d(n) = \mathbf{x}^T(n) \mathbf{w}^o(n) + v(n). \tag{4.39}$$

A first-order Markov model [22] is used for the time-varying unknown system. It has the form as follows:

$$\mathbf{w}^o(n) = \gamma \mathbf{w}^o(n-1) + \mathbf{J}(n), \tag{4.40}$$

where γ is a fixed parameter of the model and is assumed to be very close to unity. $\mathbf{J}(n)$ is the process noise vector with zero mean and correlation matrix $\mathbf{R_J}$.

The coefficient error vector is defined as

$$\mathbf{z}(n) = \mathbf{w}(n) - \mathbf{w}^o(n). \tag{4.41}$$

To determine the tracking performance of partial-update CG, three more assumptions are needed: (1) Noise $v(n)$ has zero mean and variance σ_v^2, and is independent of the noise $\mathbf{J}(n)$; (2) The input signal $\mathbf{x}(n)$ is independent of both noise $v(n)$ and noise $\mathbf{J}(n)$; (3) At steady state, the coefficient error vector $\mathbf{z}(n)$ is very small and is independent of the input signal $\mathbf{x}(n)$.

Using these assumptions, the MSE equation of the PU CG algorithm at steady state is still (4.29). At steady state, the coefficient vector is approximate to (4.30). Take the expectation of

(4.22),

$$
\begin{aligned}
E\{\widetilde{\mathbf{R}}(n)\} &= \sum_{i=1}^{n} \lambda^{n-i} E\{\hat{\mathbf{x}}(i)\mathbf{x}^{T}(i)\} \\
&= \sum_{i=1}^{n} \lambda^{n-i} \widetilde{\mathbf{R}} \\
&= \frac{\widetilde{\mathbf{R}}}{1-\lambda} \quad n \to \infty,
\end{aligned}
\tag{4.42}
$$

where $\widetilde{\mathbf{R}} = E\{\hat{\mathbf{x}}(n)\mathbf{x}^{T}(n)\}$. Assuming a slow adaptive process (λ is very close to unity), $\widetilde{\mathbf{R}}(n)$ becomes [22]

$$
\widetilde{\mathbf{R}}(n) \approx \frac{\widetilde{\mathbf{R}}}{1-\lambda} \quad n \to \infty.
\tag{4.43}
$$

The coefficient vector at steady state is further approximated to

$$
\begin{aligned}
\mathbf{w}(n) &\approx (1-\lambda)\widetilde{\mathbf{R}}^{-1}\hat{\mathbf{b}}(n) \\
&= (1-\lambda)\widetilde{\mathbf{R}}^{-1}(\lambda\hat{\mathbf{b}}(n-1) + \hat{\mathbf{x}}(n)d(n)) \\
&= \lambda\mathbf{w}(n-1) + (1-\lambda)\widetilde{\mathbf{R}}^{-1}\hat{\mathbf{x}}(n)\mathbf{x}(n)\mathbf{w}^{o}(n) + (1-\lambda)\widetilde{\mathbf{R}}^{-1}\hat{\mathbf{x}}(n)v(n).
\end{aligned}
\tag{4.44}
$$

Subtracting $\mathbf{w}^{o}(n)$ from both sides of (4.44), using (4.41) and (4.40), using the direct-averaging method [22], and applying the assumption that γ in (4.40) is very close to unity, we get

$$
\mathbf{z}(n) \approx \lambda\mathbf{z}(n-1) - \lambda\mathbf{J}(n) + (1-\lambda)\widetilde{\mathbf{R}}^{-1}\hat{\mathbf{x}}(n)v(n).
\tag{4.45}
$$

The weight error correlation matrix is (4.32). Using the property of white input noise with (4.34) and using the assumptions, $\mathbf{K}(n)$ becomes

$$
\mathbf{K}(n) \approx \mathbf{K}(n-1) + \lambda^2\mathbf{R_J} + \sigma_v^2(1-\lambda)^2 E\{\widetilde{\mathbf{R}}^{-1}\hat{\mathbf{x}}(n)\hat{\mathbf{x}}^{T}(n)\widetilde{\mathbf{R}}^{-T}\}
\tag{4.46}
$$

At steady state $\mathbf{K}(n) \approx \mathbf{K}(n-1)$, therefore $\mathbf{K}(n)$ becomes

$$
\mathbf{K}(n) \approx \frac{1-\lambda}{1+\lambda}\sigma_v^2\widetilde{\mathbf{R}}^{-1} E\{\hat{\mathbf{x}}(n)\hat{\mathbf{x}}^{T}(n)\}\widetilde{\mathbf{R}}^{-T} + \frac{\lambda^2}{1-\lambda^2}\mathbf{R_J}.
\tag{4.47}
$$

The MSE equation becomes

$$
E\{|e(n)|^2\} \approx \sigma_v^2 + tr(\mathbf{R}(\frac{1-\lambda}{1+\lambda}\sigma_v^2\widetilde{\mathbf{R}}^{-1}\hat{\mathbf{R}}\widetilde{\mathbf{R}}^{-T} + \frac{\lambda^2}{1-\lambda^2}\mathbf{R_J})),
\tag{4.48}
$$

where $tr(\cdot)$ is the trace operator, and $\hat{\mathbf{R}} = E\{\hat{\mathbf{x}}(n)\hat{\mathbf{x}}^{T}(n)\}$.

For a white input signal with variance σ_x^2, the MSE can be simplified as

$$
E\{|e(n)|^2\} \approx \sigma_v^2 + \frac{N(1-\lambda)}{1+\lambda}\sigma_v^2\sigma_x^2\sigma_{\hat{x}}^2\sigma_{\tilde{x}}^{-4} + \frac{\lambda^2}{1-\lambda^2}\sigma_x^2 tr(\mathbf{R_J}),
\tag{4.49}
$$

where $\sigma_{\hat{x}}^2 \mathbf{I} = E\{\hat{\mathbf{x}}(n)\hat{\mathbf{x}}^T(n)\}$ and $\sigma_{\underset{x}{\sim}}^{-2}\mathbf{I} = \widetilde{\mathbf{R}}^{-1}$.

For the MMax method and a white input signal, $\sigma_{\hat{x}}^2 \approx \kappa\sigma_x^2$ and $\sigma_{\underset{x}{\sim}}^2 \approx \kappa\sigma_x^2$, where κ is smaller than 1, but is close to 1. Therefore, the MSE can be further simplified as

$$E\{|e(n)|^2\} \approx \sigma_v^2 + \frac{N(1-\lambda)}{(1+\lambda)\kappa}\sigma_v^2 + \frac{\lambda^2}{1-\lambda^2}\sigma_x^2 tr(\mathbf{R_J}). \tag{4.50}$$

Assume the process noise is white with variance σ_η^2. Then, the MSE of MMax CG can be further simplified as

$$E\{|e(n)|^2\} \approx \sigma_v^2 + \frac{N(1-\lambda)}{(1+\lambda)\kappa}\sigma_v^2 + \frac{N\lambda^2}{1-\lambda^2}\sigma_x^2\sigma_\eta^2. \tag{4.51}$$

4.5   SIMULATIONS

4.5.1   PERFORMANCE OF DIFFERENT PU CG ALGORITHMS

The convergence performance and mean-square performance of different partial-update CG algorithms are compared in a system identification application. The system identification model is shown in Fig. 2.2 [5]. The unknown system [34] is a 15-order FIR filter ($N = 16$), with impulse response

$$\mathbf{w}^o = [0.01, 0.02, -0.04, -0.08, 0.15, -0.3, 0.45, 0.6,$$
$$0.6, 0.45, -0.3, 0.15, -0.08, -0.04, 0.02, 0.01]^T.$$

In our simulations, the lengths of the partial-update filter are M=8 and M=4. The variance of the input noise $v^2(n)$ is $\epsilon_{min} = 0.0001$. The initial weights are $\mathbf{w} = \mathbf{0}$. The parameters λ and η of the CG are equal to 0.99 and 0.6, respectively. The initial residue vector is set to be $\mathbf{g}(0) = d(1)\mathbf{x}(1)$. The results are obtained by averaging 100 independent runs.

The correlated input of the system [57] has the following form

$$x(n) = 0.8x(n-1) + \zeta(n), \tag{4.52}$$

where $\zeta(n)$ is zero-mean white Gaussian noise with unit variance.

Fig. 4.1 to Fig. 4.4 show the mean-square error (MSE) performance of the partial-update CG for the correlated input and white input, respectively. From the figures, we can see that the convergence rate with the white input is higher than with the correlated input and the convergence rate when PU length is 8 is higher than when the PU length is 4. The MMax CG converges a little slower than the full-update CG. The sequential and stochastic CG converge slower than the MMax CG. When PU length is 8, the steady-state MSE of the PU CG is close to the full-update CG. When PU length is 4, the steady-state MSE of the MMax CG is still close to full-update CG, while the steady-state MSE of the sequential and stochastic CG are a little higher than the full-update CG.

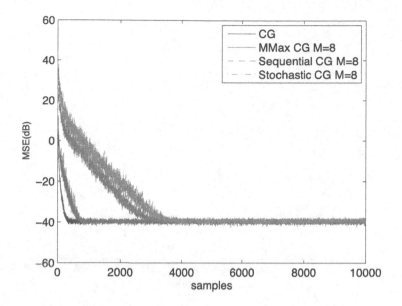

Figure 4.1: Comparison of MSE of the PU CG with correlated input, M=8.

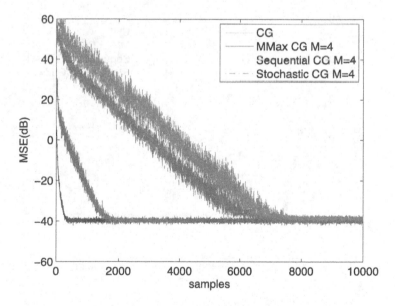

Figure 4.2: Comparison of MSE of the PU CG with correlated input, M=4.

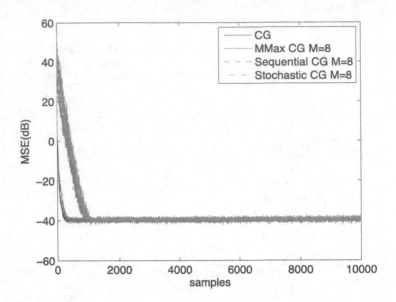

Figure 4.3: Comparison of MSE of the PU CG with white input, M=8.

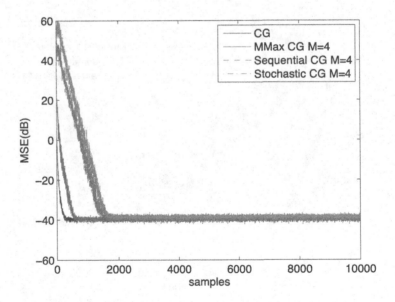

Figure 4.4: Comparison of MSE of the PU CG with white input, M=4.

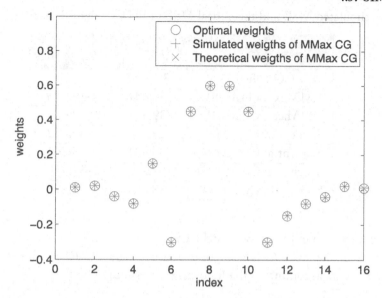

Figure 4.5: The mean convergence of the weights at steady state for MMax CG.

Fig. 4.5 shows the mean convergence of the weights at steady state for MMax CG. The PU length is 8, and the input signal is white. The theoretical results are calculated from (4.28). We can see that the simulated results match the theoretical results. The weights of MMax CG are close to the optimal weights.

Tables 4.3 and 4.4 show the simulated MSE and theoretical MSE of PU CG algorithms at steady state for correlated input and white input, respectively. The theoretical results are calculated from (4.37) and (4.38). The partial-update lengths are $M = 8$, and $M = 4$. We can see that the theoretical results match the simulated results.

Table 4.3: The simulated MSE and theoretical MSE of PU CG algorithms for correlated input

Algorithms	Simulated	Theoretical
CG (N=16)	-39.6885	–
MMax CG (M=8)	-39.65	-39.595
MMax CG (M=4)	-39.587	-39.998
Sequential CG (M=8)	-39.272	-38.767
Sequential CG (M=4)	-38.442	-38.757
Stochastic CG (M=8)	-39.324	-38.823
Stochastic CG (M=4)	-38.472	-38.779

Table 4.4: The simulated MSE and theoretical MSE of PU CG algorithms for white input

Algorithms	Simulated MSE (dB)	Theoretical MSE (dB)
CG (N=16)	-39.6885	–
MMax CG (M=8)	-39.647	-39.776
MMax CG (M=4)	-39.52	-39.664
Sequential CG (M=8)	-39.28	-39.472
Sequential CG (M=4)	-38.531	-38.861
Stochastic CG (M=8)	-39.268	-39.476
Stochastic CG (M=4)	-38.429	-38.875

Performance comparison of PU CG with PU RLS

The performance of the PU CG is also compared with the PU RLS. The PU RLS algorithm updates its coefficients according to the following recursion [11]:

$$\hat{\mathbf{P}}(0) = \delta^{-1}\mathbf{I}, \tag{4.53}$$

$$\hat{\mathbf{P}}(n) = \lambda^{-1}[\hat{\mathbf{P}}(n-1) - \frac{\lambda^{-1}\hat{\mathbf{P}}(n-1)\hat{\mathbf{x}}(n)\mathbf{x}^T(n)\hat{\mathbf{P}}(n-1)}{1 + \lambda^{-1}\mathbf{x}^T(n)\hat{\mathbf{P}}(n-1)\hat{\mathbf{x}}(n)}], \tag{4.54}$$

$$\mathbf{w}(n) = \mathbf{w}(n-1) + \hat{\mathbf{P}}(n)\hat{\mathbf{x}}(n)e(n), \tag{4.55}$$

where $\mathbf{I}$ is the identity matrix, $\hat{\mathbf{P}}(n)$ is the partial-update gain matrix and $\hat{\mathbf{P}}^{-1}(n) = \lambda^{n+1}\delta\mathbf{I} + \sum_{i=1}^{n} \lambda^{n-i}\hat{\mathbf{x}}(i)\mathbf{x}^T(i)$. δ^{-1} is an initial value. λ is the forgetting factor. The comparison uses the MMax method because the MMax method has fast convergence rate and low MSE. Fig. 4.6 shows the MSE results among CG, MMax CG, RLS, and MMax RLS. The same system identification model is used. The full-update length is 16 and the partial-update length is 8. The input signal is white. Although the full-update CG algorithm has a lower convergence rate than the full-update RLS, the MMax CG has the same convergence rate as the MMax RLS algorithm. Both MMax CG and MMax RLS can achieve similar MSE as the full-update CG and RLS at steady state. If the SORTLINE sorting method is used for both MMax CG and MMax RLS, the total number of multiplications of MMax CG and RLS are $2N^2 + M^2 + 9N + M + 3$ and $2N^2 + 2NM + 3N + M + 1$, respectively. In this case, $N = 16$ and $M = 8$. Therefore, the MMax CG needs 731 multiplications and the MMax RLS needs 825 multiplications per sample. The MMax CG needs fewer multiplications than the MMax RLS to achieve the same steady-state MSE.

Channel equalization performance is also examined among PU CG, and PU RLS algorithms. The decision-directed channel equalizer diagram is described in Fig. 2.5 [43]. A simple short FIR channel [43] is used.

$$C(z) = 0.5 + 1.2z^{-1} + 1.5z^{-2} - z^{-3}. \tag{4.56}$$

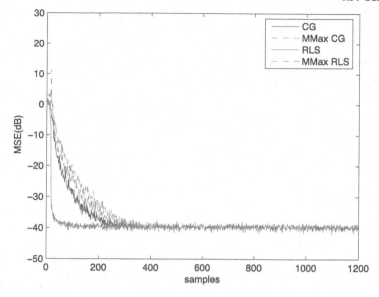

Figure 4.6: Comparison of MSE of the PU CG with the PU RLS.

The frequency response of the channel is shown in Fig. 4.7. The full length of the equalizer is assumed to be 30 and the PU length is 15. The input sequence is a 4-QAM signal. The results are obtained by averaging 50 independent runs. Fig. 4.8 illustrates the symbol-error-rate (SER) in log-scale among CG, MMax CG, RLS, and MMax RLS algorithms. The SER performance of these algorithms are still related to the MSE performance shown in Fig. 4.6. The full-update CG and RLS have similar SER performance. The MMax CG and MMax RLS have similar performance, and their performance is also close to the full-update algorithms.

4.5.2 TRACKING PERFORMANCE OF THE PU CG USING THE FIRST-ORDER MARKOV MODEL

The system identification model is shown in Fig. 2.2. The initial optimal weights are still

$$\mathbf{w}^o(0) = [0.01, 0.02, -0.04, -0.08, 0.15, -0.3, 0.45, 0.6,$$
$$0.6, 0.45, -0.3, 0.15, -0.08, -0.04, 0.02, 0.01]^T. \tag{4.57}$$

In our simulations, the lengths of the partial update filter are M=8 and M=4. The variance of the input noise $v^2(n)$ is 0.0001. The initial weights are $\mathbf{w} = \mathbf{0}$. The parameters λ and η of the CG are equal to 0.99 and 0.6, respectively. The initial residue vector is set to be $\mathbf{g}(0) = d(1)\mathbf{x}(1)$. The results are obtained by averaging 100 independent runs. The white process noise is used with difference variances. The white input signal with unity variance is used.

Fig. 4.9 shows the tracking performance of the PU CG with process noise $\sigma_\eta = 0.001$ and PU length $M = 8$. Fig. 4.10 shows the tracking performance of PU CG with process noise

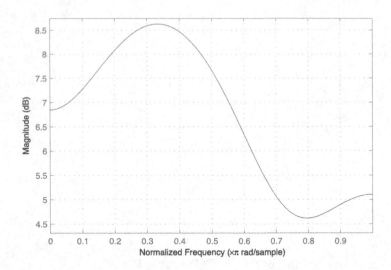

Figure 4.7: The frequency response of the channel.

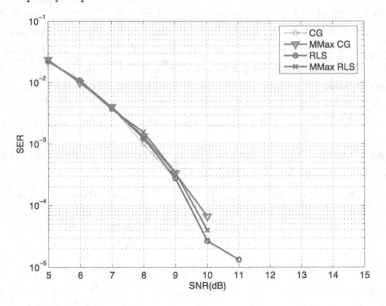

Figure 4.8: Comparison of SER of the PU CG with the PU RLS.

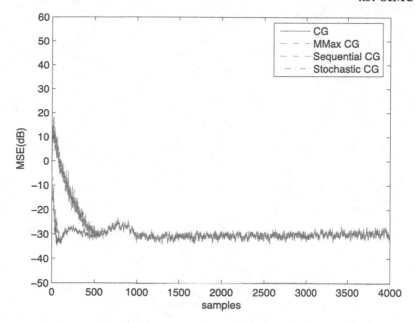

Figure 4.9: Comparison of MSE of the PU CG with CG for white input, N=16, M=8, and $\sigma_\eta = 0.001$.

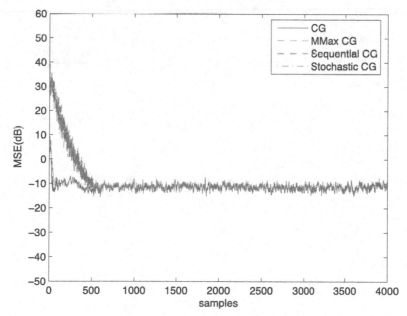

Figure 4.10: Comparison of MSE of the PU CG with CG for white input, N=16, M=8, and $\sigma_\eta = 0.01$.

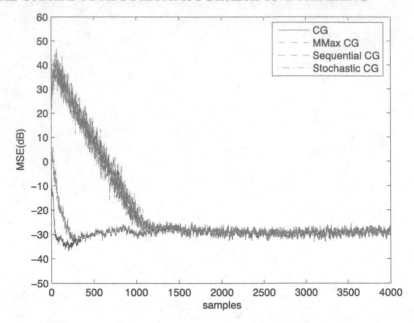

Figure 4.11: Comparison of MSE of the PU CG with CG for white input, N=16, M=4, and $\sigma_\eta = 0.001$.

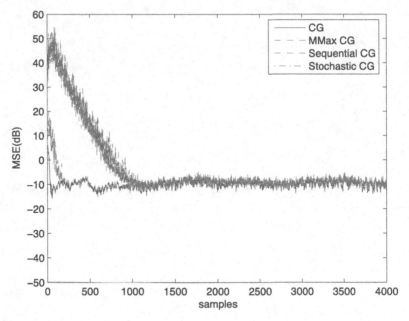

Figure 4.12: Comparison of MSE of the PU CG with CG for white input, N=16, M=4, and $\sigma_\eta = 0.01$.

$\sigma_\eta = 0.01$ and the PU length $M = 8$. All PU CGs have similar steady-state MSE performance. We can see that the MSE of the PU CG increases when the process noise increases. The variance of the MSE also increases when the process noise increases. Again, this is because the temporal MSE increases as the non-stationary degree increases. The same situation also happens to the full-update CG. Fig. 4.11 shows the tracking performance of the PU CG with process noise $\sigma_\eta = 0.001$ and PU length $M = 4$. Fig. 4.12 shows the tracking performance of the PU CG with process noise $\sigma_\eta = 0.01$ and PU length $M = 4$. The sequential and stochastic CG perform a little worse than the MMax CG and full-update CG at steady state. The PU length does not affect the steady-state MSE too much in this case. The PU length affects the convergence rate. The convergence rate decreases as the partial update length decreases.

Table 4.5 and Table 4.6 show the simulated MSE and theoretical MSE of PU CG algorithms at steady state for white input for process noise $\sigma_\eta = 0.001$ and $\sigma_\eta = 0.01$, respectively. The simulated results are obtained by taking the time average over the last 1,000 samples. The theoretical results are calculated from (4.51). The partial-update lengths are $M = 8$ and $M = 4$. We can see that the theoretical results match the simulated results.

Table 4.5: The simulated MSE and theoretical MSE of the PU CG in a time-varying system for white input and process noise $\sigma_\eta = 0.001$

Algorithms	Simulated MSE (dB)	Theoretical MSE (dB)
CG (N=16)	-29.81	-30.523
MMax CG (M=8)	-29.745	-30.519
Sequential CG (M=8)	-29.375	-30.484
Stochastic CG (M=8)	-29.377	-30.484
MMax CG (M=4)	-29.9034	-30.437
Sequential CG (M=4)	-28.8422	-30.339
Stochastic CG (M=4)	-28.6976	-30.339

Performance Comparison of the MMax CG with the CG, RLS, and MMax RLS for a Time-Varying System

The tracking performance of the MMax CG is also compared with the full-update CG, full-update RLS, and MMax RLS. The same system identification model is used. After 2,000 samples/iterations pass, the unknown system in (4.57) is changed by multiplying all coefficients by -1. Fig. fig: CGMMaxCGRLSMmaxRLSwhite_M8 and Fig. 4.14 show the MSE results among CG, MMax CG, RLS, and MMax RLS, when $M = 8$ and $M = 4$, respectively. The white input is used. The results show that the four algorithms have a similar convergence rate after the unknown system changes. The initial convergence rates of the four algorithms are different because there is no memory for $\mathbf{R}(n)$ or $\hat{\mathbf{R}}(n)$ at the initial state. The PU autocorrelation matrix $\hat{\mathbf{R}}(n)$ be-

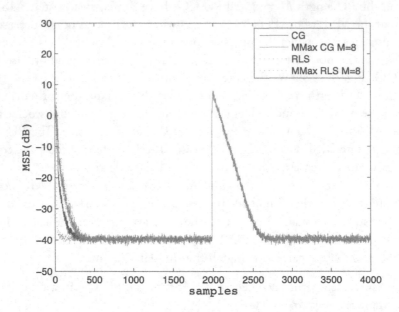

Figure 4.13: Comparison of MSE of the MMax CG with CG, RLS, and MMax RLS for white input, N=16, M=8.

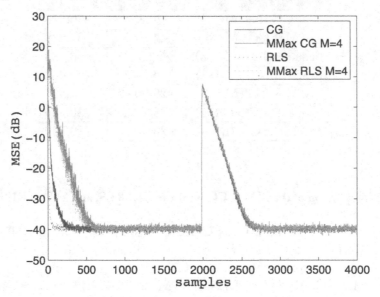

Figure 4.14: Comparison of MSE of the MMax CG with CG, RLS, and MMax RLS for white input, N=16, M=4.

Table 4.6: The simulated MSE and theoretical MSE of the PU CG in a time-varying system for white input and process noise $\sigma_\eta = 0.01$

Algorithms	Simulated MSE (dB)	Theoretical MSE (dB)
CG (N=16)	-11.172	-10.948
MMax CG (M=8)	-11.14	-10.948
Sequential CG (M=8)	-10.683	-10.747
Stochastic CG (M=8)	-10.768	-10.747
MMax CG (M=4)	-9.9862	-9.967
Sequential CG (M=4)	-8.7336	-9.966
Stochastic CG (M=4)	-8.6443	-9.966

comes close to $\mathbf{R}(n)$ after a few updates. The previous input data are still memorized by the filters. When the unknown system changes, $\mathbf{R}(n)$ and $\hat{\mathbf{R}}(n)$ still have the wrong memory, and therefore the convergence rates of the four algorithms are similar and slower than the initial convergence rates. The MMax CG and MMax RLS with $M = 4$ also have similar convergence rates to the MMax CG and MMax RLS with $M = 8$ after the unknown system change. If the SORTLINE sorting method is used for both MMax CG and MMax RLS, the total number of multiplications of MMax CG and RLS are $2N^2 + M^2 + 9N + M + 3$ and $2N^2 + 2NM + 3N + M + 1$, respectively. In this case, the full update length N is 16. The partial-update length M is 8 and 4, respectively. The detailed computational complexities of the four algorithms are shown in Table 4.7. The results show that the MMax CG with M=4 can achieve similar tracking performance to the full-update RLS or CG while reducing the computational complexity significantly.

Table 4.7: The computational complexities of CG, MMax CG, RLS, and MMax RLS

Algorithms	Number of multiplications per symbol	Number of comparisons per symbol
CG (N=16)	3003	–
MMax CG (M=8)	731	10
MMax CG (M=4)	679	10
RLS (N=16)	3721	–
MMax RLS (M=8)	825	10
MMax RLS (M=4)	693	10

4.6   CONCLUSION

In this chapter, different PU CG algorithms were developed. Theoretical mean and mean-square performance were derived for both time-invariant and time-varying systems. The performance of different PU CG algorithms was compared by using computer simulations. The simulated results matched the theoretical results. We can conclude that the PU CG can achieve similar steady-state MSE as the full-update CG while reducing the computational complexity significantly. Among different PU CG algorithms, the MMax CG algorithm has a convergence rate close to the full-update CG. The performance of the PU CG was also compared with the PU RLS by using the MMax method. The MMax CG has a similar performance to the MMax RLS, while having a lower computational complexity.

CHAPTER 5

Partial-Update EDS Algorithms for Adaptive Filtering

In this chapter, partial-update techniques are applied to the Euclidean Direction Search (EDS) algorithm. The theoretical mean and mean-square expressions of the PU EDS in a time-invariant system are derived. The performance of the PU EDS is analyzed using a system identification model and a channel equalizer system. The theoretical mean-square expressions of the PU EDS in a time-varying system are also derived. The tracking performance of the PU EDS is analyzed using a system identification model.

5.1 MOTIVATION

The EDS algorithm, developed by Xu and Bose [55], has similar characteristics to the RLS algorithm. It also solves the least squares cost function, but uses a direction search method. It has a fast convergence rate and small mean-square error. However, the high computational complexity $O(N^2)$ makes it unsuitable for many real-time processing. Partial-update methods are applied to the EDS algorithm.

5.2 REVIEW OF EUCLIDEAN DIRECTION SEARCH ALGORITHM

The Euclidean Direction Search Algorithm solves a least squares cost function:

$$J(n) = \sum_{i=1}^{n} \lambda^{n-i} e^2(n) \tag{5.1}$$

$$= \sum_{i=1}^{n} \lambda^{n-i} [d(i) - y(i)]^2, \tag{5.2}$$

where $y(i) = \mathbf{x}^T(i)\mathbf{w}(i)$.

The original EDS algorithm updates its coefficients according to the following recursion [5]:

$$\mathbf{w}(n+1) = \mathbf{w}(n) + \alpha\mathbf{g}, \tag{5.3}$$

$$\alpha = -\frac{\mathbf{g}^T(\mathbf{Q}(n)\mathbf{w}(n) - \mathbf{r}(n))}{\mathbf{g}^T\mathbf{Q}(n)\mathbf{g}}, \tag{5.4}$$

$$\mathbf{Q}(n) = \sum_{i=1}^{n}\lambda^{n-i}\mathbf{x}(i)\mathbf{x}^T(i), \tag{5.5}$$

$$\mathbf{r}(n) = \sum_{i=1}^{n}\lambda^{n-i}d(i)\mathbf{x}(i), \tag{5.6}$$

where the $N \times 1$ vector $\mathbf{g}$ is the search direction at iteration n, which is taken to be "Euclidean directions." It is defined as $\mathbf{g}_i = [0, \cdots, 0, 1, 0, \cdots, 0]^T$, where the 1 appears in the i-th position. At each iteration n, the entire weight vector $\mathbf{w}(n)$ is updated by cycling through all the Euclidean directions $\mathbf{g}_i, i = 1, 2, \cdots, N$ [57]. The total number of multiplications needed for each incoming data sample (for real signals) is $N(3N + 3)$ [5].

5.3   PARTIAL UPDATE EDS

Basic partial update methods will be applied to the EDS algorithm including periodic, sequential, stochastic, the MMax update method.

The partial update EDS has the uniform update equation:

$$\mathbf{w}(n+1) = \mathbf{w}(n) - \frac{\mathbf{g}\mathbf{g}^T(\widetilde{\mathbf{Q}}(n)\mathbf{w}(n) - \hat{\mathbf{r}}(n))}{\mathbf{g}^T\widetilde{\mathbf{Q}}(n)\mathbf{g}}, \tag{5.7}$$

where

$$\widetilde{\mathbf{Q}}(n) = \sum_{i=1}^{n}\lambda^{n-i}\mathbf{x}(i)\hat{\mathbf{x}}^T(i), \tag{5.8}$$

$$\hat{\mathbf{r}}(n) = \sum_{i=1}^{n}\lambda^{n-i}\hat{\mathbf{x}}(i)d(i), \tag{5.9}$$

$$\hat{\mathbf{x}} = \mathbf{I}_M\mathbf{x}, \tag{5.10}$$

$$\mathbf{I}_M(n) = \begin{bmatrix} i_1(n) & 0 & \dots & 0 \\ 0 & i_2(n) & \ddots & \vdots \\ \vdots & \ddots & \ddots & 0 \\ 0 & \dots & 0 & i_N(n) \end{bmatrix}, \tag{5.11}$$

$$\sum_{k=1}^{N} i_k(n) = M, \quad i_k(n) \in \{0, 1\}, \tag{5.12}$$

The partial update subset $\mathcal{I}_M(n)$ has been defined in Section 2.4.

For PU EDS, the entire weight vector $\mathbf{w}(n)$ is partially updated by cycling through the Euclidean directions $\mathbf{g}_i$ at each iteration n if $i \in \mathcal{I}_M(n)$. This means if $i \in \mathcal{I}_M(n)$, then the ith element of $\mathbf{g}$ will be set to be 1 and the ith element of $\mathbf{w}(n)$ will be updated. Otherwise, the ith element of $\mathbf{w}(n)$ will not be updated. The sorting of the input $\mathbf{x}$ increases the computational complexity. The sorting result can be achieved more efficiently by using the SORTLINE or Shortsort methods [7].

The detailed computational complexities of different PU EDS are shown in Table 5.1 and Table 5.2. The SORTLINE method is used and the MMax EDS needs $2 + 2 \lceil \log_2 N \rceil$ comparisons. Table 5.1 shows the computational complexity of different PU EDS for real-valued data in terms of the number of real multiplications and real additions. Table 5.2 shows the computational complexity of different PU EDS for complex-valued data in terms of the number of real multiplications and real additions. If M is much smaller than N, then the number of multiplications can be reduced significantly for PU EDS.

Table 5.1: The computational complexities of different PU EDS for real-valued data in terms of the number of real multiplications and real additions

Algorithms	Number of multiplications per sample	Number of additions per sample	Comparisons
EDS	$3N^2 + 3N$	$2N^2 + 2N$	0
Sequential EDS	$N^2 + 2NM + N + 2M$	$2NM + 2M$	0
Stochastic EDS	$N^2 + 2NM + N + 2M + 2$	$2NM + 2M + 2$	0
MMax EDS	$N^2 + 2NM + N + 2M$	$2NM + 2M$	$2 [log_2 N] + 2$

Table 5.2: The computational complexities of different PU EDS for complex-valued data in terms of the number of real multiplications and real additions

Algorithms	Number of multiplications per sample	Number of additions per sample	Comparisons
EDS	$10N^2 + 10N$	$10N^2 + 10N$	0
Sequential EDS	$6N^2 + 4NM + 6N + 4M$	$5N^2 + 5NM + 7N + 3M$	0
Stochastic EDS	$6N^2 + 4NM + 6N + 4M + 2$	$5N^2 + 5NM + 7N + 3M + 2$	0
MMax EDS	$6N^2 + 4NM + 8N + 4M$	$5N^2 + 5NM \, C \, 8N + 3M$	$2 [log_2 N] + 2$

5.4   PERFORMANCE OF THE PARTIAL-UPDATE EDS IN A TIME-INVARIANT SYSTEM

The mean behavior of the PU EDS weights can be determined by multiplying a scalar $\mathbf{g}^T \widetilde{\mathbf{Q}}(n)\mathbf{g}$ to both sides of (5.7) and taking the expectation

$$
\begin{aligned}
E\{\mathbf{g}^T \widetilde{\mathbf{Q}}(n)\mathbf{g}\mathbf{w}(n+1)\} &= E\{\mathbf{g}^T \widetilde{\mathbf{Q}}(n)\mathbf{g}\mathbf{w}(n)\} \\
&- E\{\mathbf{g}\mathbf{g}^T (\widetilde{\mathbf{Q}}(n)\mathbf{w}(n) - \hat{\mathbf{r}}(n))\}.
\end{aligned}
\tag{5.13}
$$

Assume $\widetilde{\mathbf{Q}}(n)$ and $\mathbf{w}(n)$ are uncorrelated to each other. $\mathbf{g}\mathbf{g}^T$ is just direction and is uncorrelated to $\widetilde{\mathbf{Q}}(n)$ and $\hat{\mathbf{r}}(n)$. At steady state, $E\{\mathbf{g}^T \widetilde{\mathbf{Q}}(n)\mathbf{g}\mathbf{w}(n+1)\} = E\{\mathbf{g}^T \widetilde{\mathbf{Q}}(n)\mathbf{g}\mathbf{w}(n)\}$. Therefore, (5.13) can be simplified to

$$
E\{\mathbf{g}\mathbf{g}^T\}E\{\widetilde{\mathbf{Q}}(n)\}E\{\mathbf{w}(n)\} = E\{\mathbf{g}\mathbf{g}^T\}E\{\hat{\mathbf{r}}(n)\}.
\tag{5.14}
$$

At steady state, $E\{\widetilde{\mathbf{Q}}(n)\} = \frac{1}{1-\lambda}\widetilde{\mathbf{Q}}$ and $E\{\hat{\mathbf{r}}(n)\} = \frac{1}{1-\lambda}\hat{\mathbf{r}}$, where $\widetilde{\mathbf{Q}} = E\{\mathbf{x}(n)\hat{\mathbf{x}}^T(n)\}$ and $\hat{\mathbf{r}} = E\{\hat{\mathbf{x}}(n)d(n)\}$. If the inversion of $\widetilde{\mathbf{Q}}$ exists, the mean weights of the PU EDS converge to

$$
E\{\mathbf{w}(n)\} = \widetilde{\mathbf{Q}}^{-1}\hat{\mathbf{r}} \quad n \to \infty.
\tag{5.15}
$$

The choice of PU filter length M needs to satisfy the condition that the inversion of $\widetilde{\mathbf{Q}}$ exists. For the MMax method $\widetilde{\mathbf{Q}}$ is close to that of the full-update method. For the sequential and stochastic methods, $\widetilde{\mathbf{Q}}$ is different from that of the full-update method. The inversion of $\widetilde{\mathbf{Q}}$ does not always exist, especially when the PU length is small.

The coefficient error vector is defined as

$$
\mathbf{z}(n) = \mathbf{w}(n) - \mathbf{w}^o.
\tag{5.16}
$$

To derive the MSE performance at steady state, three assumptions are needed: (1) Inversion of $\widetilde{\mathbf{Q}}$ exists; (2) at steady state, the coefficient error vector $\mathbf{z}(n)$ is very small and is independent of the input signal $\mathbf{x}(n)$; (3) the input signal $\mathbf{x}(n)$ is independent of noise $v(n)$.

Define the weight error correlation matrix as

$$
\mathbf{K}(n) = E\{\mathbf{z}(n)\mathbf{z}^T(n)\}.
\tag{5.17}
$$

Using the assumptions, the MSE equation of the PU EDS algorithm at steady state becomes

$$
E\{|e(n)|^2\} = \sigma_v^2 + tr(\mathbf{Q}\mathbf{K}(n)),
\tag{5.18}
$$

where $\mathbf{Q}$ is the autocorrelation matrix of the input $\mathbf{x}$.

At steady state, the coefficient vector is approximate to

$$
\mathbf{w}(n) \approx \widetilde{\mathbf{Q}}^{-1}(n)\hat{\mathbf{r}}(n).
\tag{5.19}
$$

Assuming a slow adaptive process (λ is very close to unity), $\widetilde{\mathbf{Q}}(n)$ becomes [22]

$$\widetilde{\mathbf{Q}}(n) \approx \frac{\widetilde{\mathbf{Q}}}{1-\lambda} \quad n \to \infty. \tag{5.20}$$

The coefficient vector is further approximated to

$$\begin{aligned}
\mathbf{w}(n) &\approx (1-\lambda)\widetilde{\mathbf{Q}}^{-1}\hat{\mathbf{r}}(n) \\
&= \lambda\mathbf{w}(n-1) + (1-\lambda)\widetilde{\mathbf{Q}}^{-1}\hat{\mathbf{x}}(n)\mathbf{x}^T(n)\mathbf{w}^o \\
&+ (1-\lambda)\widetilde{\mathbf{Q}}^{-1}\hat{\mathbf{x}}(n)\mathbf{v}(n).
\end{aligned} \tag{5.21}$$

Subtracting $\mathbf{w}^o$ from both sides of (5.21), using (5.16) and the direct-averaging method [22], we get

$$\mathbf{z}(n) \approx \lambda\mathbf{z}(n-1) + (1-\lambda)\widetilde{\mathbf{Q}}^{-1}\hat{\mathbf{x}}(n)\mathbf{v}(n). \tag{5.22}$$

Note, after using the direct-averaging method, the term $(1-\lambda)\widetilde{\mathbf{Q}}^{-1}\hat{\mathbf{x}}(n)\mathbf{x}^T(n)\mathbf{w}^o$ in (5.21) becomes $(1-\lambda)\widetilde{\mathbf{Q}}^{-1}E\{\hat{\mathbf{x}}(n)\mathbf{x}^T(n)\}\mathbf{w}^o = (1-\lambda)\mathbf{w}^o$.

Since the input noise is assumed to be white,

$$E\{v(i)v(j)\} = \begin{cases} \sigma_v^2 & \text{for } i = j \\ 0 & \text{otherwise} \end{cases}. \tag{5.23}$$

$\mathbf{K}(n)$ becomes

$$\mathbf{K}(n) \approx \lambda^2\mathbf{K}(n-1) + \sigma_v^2(1-\lambda)^2 E\{\widetilde{\mathbf{Q}}^{-1}\hat{\mathbf{x}}(n)\hat{\mathbf{x}}^T(n)\widetilde{\mathbf{Q}}^{-T}\} \tag{5.24}$$

At steady state $\mathbf{K}(n) \approx \mathbf{K}(n-1)$, therefore $\mathbf{K}(n)$ becomes

$$\mathbf{K}(n) \approx \frac{1-\lambda}{1+\lambda}\sigma_v^2\widetilde{\mathbf{Q}}^{-1}E\{\hat{\mathbf{x}}(n)\hat{\mathbf{x}}^T(n)\}\widetilde{\mathbf{Q}}^{-T}. \tag{5.25}$$

The MSE equation becomes

$$E\{|e(n)|^2\} \approx \sigma_v^2 + tr(\mathbf{Q}(\frac{1-\lambda}{1+\lambda}\sigma_v^2\widetilde{\mathbf{Q}}^{-1}\hat{\mathbf{Q}}\widetilde{\mathbf{Q}}^{-T})), \tag{5.26}$$

where $tr(\cdot)$ is the trace operator and $\hat{\mathbf{Q}} = E\{\hat{\mathbf{x}}(n)\hat{\mathbf{x}}^T(n)\}$.

For a white input signal with variance σ_x^2, the MSE can be simplified as

$$E\{|e(n)|^2\} \approx \sigma_v^2 + \frac{N(1-\lambda)}{1+\lambda}\sigma_v^2\sigma_x^2\sigma_{\hat{x}}^2\sigma_{\tilde{x}}^{-4}, \tag{5.27}$$

where $\sigma_{\hat{x}}^2\mathbf{I} = E\{\hat{\mathbf{x}}(n)\hat{\mathbf{x}}^T(n)\}$ and $\sigma_{\tilde{x}}^{-2}\mathbf{I} = \widetilde{\mathbf{Q}}^{-1}$.

For the PU method and a white input signal, $\sigma_{\hat{x}}^2 \approx \kappa\sigma_x^2$ and $\sigma_{\tilde{x}}^2 \approx \kappa\sigma_x^2$, where κ is smaller than 1 and is close to 1. Therefore, the MSE can be further simplified as

$$E\{|e(n)|^2\} \approx \sigma_v^2 + \frac{N(1-\lambda)}{(1+\lambda)\kappa}\sigma_v^2. \tag{5.28}$$

5.5   PERFORMANCE OF THE PARTIAL-UPDATE EDS IN A TIME-VARYING SYSTEM

In a non-stationary environment, the unknown system is time-varying. The desired signal can be rewritten as

$$d(n) = \mathbf{x}^T(n)\mathbf{w}^o(n) + v(n). \tag{5.29}$$

A first-order Markov model [22] is used for the time-varying unknown system. It has the form

$$\mathbf{w}^o(n) = \gamma \mathbf{w}^o(n-1) + \mathbf{j}(n), \tag{5.30}$$

where γ is a fixed parameter of the model and is assumed to be very close to unity. $\mathbf{j}(n)$ is the process noise vector with zero mean and correlation matrix $\mathbf{Q_j}$.

The coefficient error vector for the time-varying system is defined as

$$\mathbf{z}(n) = \mathbf{w}(n) - \mathbf{w}^o(n). \tag{5.31}$$

To determine the tracking performance of the partial update EDS, two more assumptions are needed [22]: (1) Noise $v(n)$ has zero mean and variance σ_v^2, and is independent of the noise $\mathbf{j}(n)$; (2) the input signal $\mathbf{x}(n)$ is independent of both noise $v(n)$ and noise $\mathbf{j}(n)$.

Using the same derivation steps as the time-invariant system, the coefficient vector at steady state is approximated to

$$\begin{aligned}\mathbf{w}(n) &\approx \lambda \mathbf{w}(n-1) + (1-\lambda)\widetilde{\mathbf{Q}}^{-1}\hat{\mathbf{x}}(n)\mathbf{x}^T(n)\mathbf{w}^o(n) \\ &+ (1-\lambda)\widetilde{\mathbf{Q}}^{-1}\hat{\mathbf{x}}(n)\mathbf{v}(n).\end{aligned} \tag{5.32}$$

Subtracting $\mathbf{w}^o(n)$ from both sides of (5.32), using (5.30) and (5.31), using the direct-averaging method [22], and applying the assumption that γ in (5.30) is very close to unity, we obtain

$$\mathbf{z}(n) \approx \lambda \mathbf{z}(n-1) - \lambda \mathbf{j}(n) + (1-\lambda)\widetilde{\mathbf{Q}}^{-1}\hat{\mathbf{x}}(n)\mathbf{v}(n). \tag{5.33}$$

$\mathbf{K}(n)$ becomes

$$\mathbf{K}(n) \approx \lambda^2 \mathbf{K}(n-1) + \lambda^2 \mathbf{Q_j} + \sigma_v^2(1-\lambda)^2 E\{\widetilde{\mathbf{Q}}^{-1}\hat{\mathbf{x}}(n)\hat{\mathbf{x}}^T(n)\widetilde{\mathbf{Q}}^{-T}\} \tag{5.34}$$

At steady state $\mathbf{K}(n) \approx \mathbf{K}(n-1)$, therefore $\mathbf{K}(n)$ becomes

$$\mathbf{K}(n) \approx \frac{1-\lambda}{1+\lambda}\sigma_v^2 \widetilde{\mathbf{Q}}^{-1} E\{\hat{\mathbf{x}}(n)\hat{\mathbf{x}}^T(n)\}\widetilde{\mathbf{Q}}^{-T} + \frac{\lambda^2}{1-\lambda^2}\mathbf{Q_j}. \tag{5.35}$$

The MSE equation becomes

$$E\{|e(n)|^2\} \approx \sigma_v^2 + tr(\mathbf{Q}(\frac{1-\lambda}{1+\lambda}\sigma_v^2 \widetilde{\mathbf{Q}}^{-1}\hat{\mathbf{Q}}\widetilde{\mathbf{Q}}^{-T} + \frac{\lambda^2}{1-\lambda^2}\mathbf{Q_j})). \tag{5.36}$$

For a white input signal with variance σ_x^2, the MSE can be simplified as

$$E\{|e(n)|^2\} \quad \approx \quad \sigma_v^2 + \frac{N(1-\lambda)}{1+\lambda}\sigma_v^2\sigma_x^2\sigma_{\hat{x}}^2\tilde{\sigma}_x^{-4} + \frac{\lambda^2}{1-\lambda^2}\sigma_x^2 tr(\mathbf{Q_J}). \tag{5.37}$$

For the PU method and a white input signal, the MSE can be further simplified as

$$E\{|e(n)|^2\} \approx \sigma_v^2 + \frac{N(1-\lambda)}{(1+\lambda)\kappa}\sigma_v^2 + \frac{\lambda^2}{1-\lambda^2}\sigma_x^2 tr(\mathbf{Q_J}). \tag{5.38}$$

Assume the process noise is white with variance σ_η^2. Then, the MSE of the PU EDS can be further simplified as

$$E\{|e(n)|^2\} \approx \sigma_v^2 + \frac{N(1-\lambda)}{(1+\lambda)\kappa}\sigma_v^2 + \frac{N\lambda^2}{1-\lambda^2}\sigma_x^2\sigma_\eta^2. \tag{5.39}$$

5.6   SIMULATIONS

5.6.1   PERFORMANCE OF THE PU EDS IN A TIME-INVARIANT SYSTEM

The system identification model shown in Fig. 2.2 is used. It is a 15th-order FIR filter (N=16). The impulse response [34] is

$$\mathbf{w}^o = [0.01, 0.02, -0.04, -0.08, 0.15, -0.3, 0.45, 0.6,$$
$$0.6, 0.45, -0.3, 0.15, -0.08, -0.04, 0.02, 0.01]^T. \tag{5.40}$$

In our simulations, the lengths of the partial update filter are $M = 8$ and $M = 4$. The variance of the input noise $v(n)$ is 0.0001. The initial weights of the EDS are $\mathbf{w} = \mathbf{0}$, the initial autocorrelation matrix $\mathbf{Q}(0) = \mathbf{0}$, and the crosscorrelation vector $\mathbf{r}(0) = \mathbf{0}$. The parameter λ is equal to 0.99. The results are obtained by averaging 100 independent runs. The number of input samples used for each run is 2,000. To show the convergence performance clearly, only 1,200 samples are plotted in the figures which show the MSE performance. When calculating the simulation results in the tables which compare the simulated steady-state MSE and theoretical steady-state MSE, the last 1,000 samples (of 2,000 samples) are used. Both correlated input and white input are used. The correlated input of the system [57] has the following form

$$x(n) = 0.8x(n-1) + \beta(n), \tag{5.41}$$

where $\beta(n)$ was zero-mean white Gaussian noise with unit variance.

Figure 5.1 and Fig. 5.2 show the MSE performance of the PU EDS in a time-invariant system with white input, and for PU length $M = 8$ and $M = 4$, respectively. We can see that all the PU EDS algorithms can converge to the same steady-state MSE as the full-update EDS for $M = 8$. The MMax EDS has a converge rate close to the full-update EDS. The sequential and stochastic methods have slightly higher steady-state MSE than the full-update EDS when

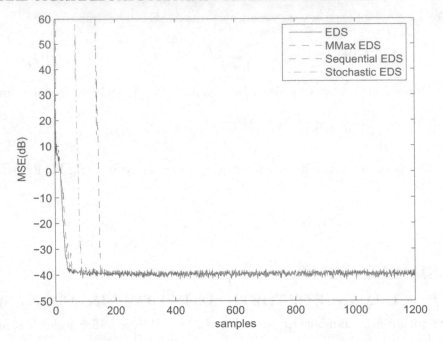

Figure 5.1: Comparison of MSE of PU EDS with white input, $M = 8$.

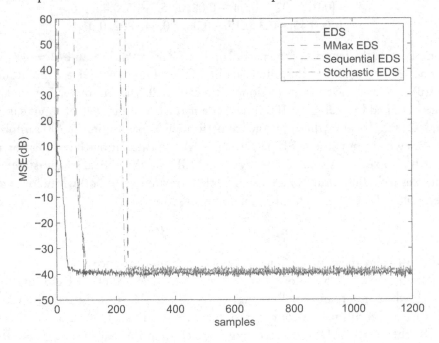

Figure 5.2: Comparison of MSE of PU EDS with white input, $M = 4$.

the PU length is $M = 4$. The MMax EDS has a similar steady-state MSE to the full-update EDS when $M = 4$. The convergence rate of PU length $M = 4$ is slower than that of PU length $M = 8$.

Table 5.3 shows the simulated MSE and theoretical MSE of PU EDS algorithms at steady state for white input. The simulated results are obtained by taking the time average over the last 1,000 samples. The theoretical results are calculated from (5.28). For full-update EDS, $\kappa = 1$. We can see that the simulated results match the theoretical results.

Table 5.3: The simulated MSE and theoretical MSE of PU EDS for time-invariant system and white input

Algorithms	Simulated MSE (dB)	Theoretical MSE (dB)
EDS (N=16)	-39.662	-39.664
MMax EDS (N=8)	-39.625	-39.606
Sequential EDS (N=8)	-39.284	-38.807
Stochastic EDS (N=8)	-39.295	-38.812
MMax EDS (N=4)	-39.500	-39.316
Sequential EDS (N=4)	-38.386	-36.364
Stochastic EDS (N=4)	-38.379	-36.364

Figure 5.3 shows the MSE performance of the PU EDS in a time-invariant system with correlated input vector **x**, and for PU length $M = 8$. We can see that the MMax EDS still has the best performance among the different PU EDS algorithms. It can converge the same steady-state MSE as the full-update EDS. The convergence rate is also close to the full-update EDS. Figure 5.4 shows the MSE performance of the PU EDS in a time-invariant system with correlated input vector **x**, and for PU length $M = 4$. The PU EDS are not stable with PU length $M - 4$.

Table 5.4 shows the simulated MSE and theoretical MSE of PU EDS algorithms at steady state for correlated input at PU length $M = 8$. Since the filter is not stable for correlated input at PU length $M = 4$, the MSE result for $M = 4$ is not shown. The simulated results are obtained by taking the time average over the last 1,000 samples. The theoretical results are calculated from (5.26). The theoretical results match the simulated results for correlated input.

Table 5.4: The simulated MSE and theoretical MSE of the PU EDS for a time-invariant system and correlated input

Algorithms	Simulated MSE (dB)	Theoretical MSE (dB)
EDS (N=16)	-39.650	-39.664
MMax EDS (N=8)	-39.516	-39.546
Sequential EDS (N=8)	-39.059	-39.220
Stochastic EDS (N=8)	-39.074	-39.221

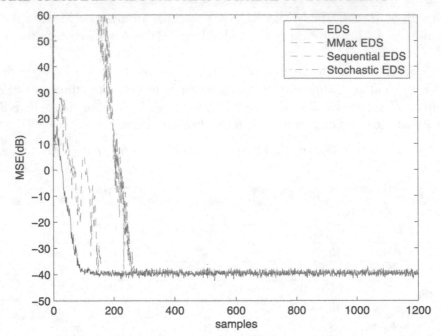

Figure 5.3: Comparison of MSE of PU EDS with correlated input, $M = 8$.

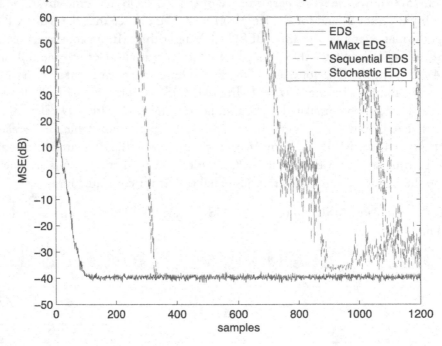

Figure 5.4: Comparison of MSE of PU EDS with correlated input, $M = 4$.

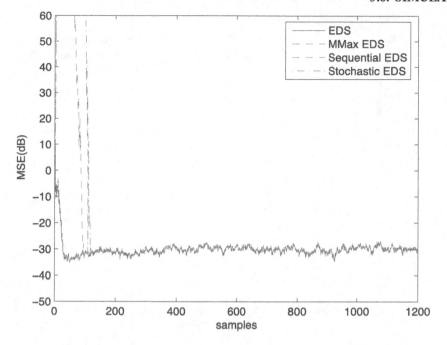

Figure 5.5: Comparison of MSE of PU EDS with EDS for white input, $N = 16$, $M = 8$, $\sigma_\eta = 0.001$.

5.6.2   TRACKING PERFORMANCE OF THE PU EDS USING THE FIRST-ORDER MARKOV MODEL

The same system identification model for a time-invariant system is used, except the weights are time-varying. The first-order Markov model (5.30) is used for the time-varying impulse response. The initial state of the impulse response is (5.40). The parameter γ in the first Markov model is 0.9998. The white process noise is used with difference variances. The white input signal with unity variance is used.

Figure 5.5 and Fig. 5.6 show the tracking performance of the PU EDS with a different process noise $\sigma_\eta = 0.001$ and $\sigma_\eta = 0.01$ for $M = 8$. All PU EDS have similar steady-state MSE performance. We can see that the MSE of PU EDS increases when the process noise increases. The variance of the MSE also increases when the process noise increases. The same situation also happens to the full-update EDS. Figure 5.7 and Figure 5.8 show the tracking performance of the PU EDS with a different process noise $\sigma_\eta = 0.001$ and $\sigma_\eta = 0.01$ for $M = 4$. The partial-update length does not have much effect on the MSE results in this case. The partial-update length only affects the convergence rate. The convergence rate decreases as the partial-update length decreases.

Table 5.5 and Table 5.6 show the simulated MSE and theoretical MSE of PU EDS algorithms at steady state for white input for process noise $\sigma_\eta = 0.001$ and $\sigma_\eta = 0.01$, respectively.

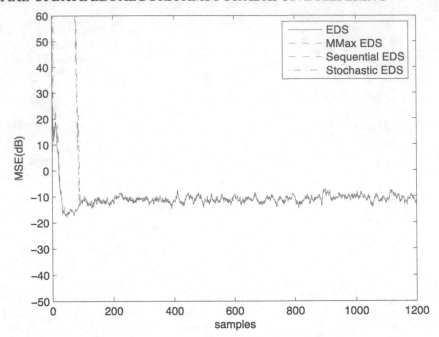

Figure 5.6: Comparison of MSE of PU EDS with EDS for white input, $N = 16$, $M = 8$, $\sigma_\eta = 0.01$.

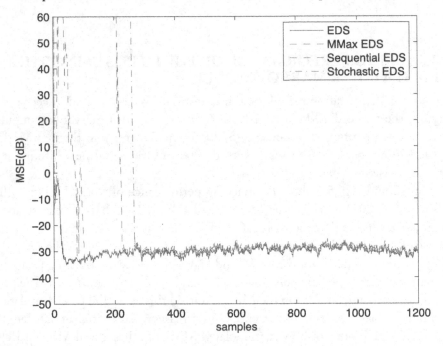

Figure 5.7: Comparison of MSE of PU EDS with EDS for white input, $N = 16$, $M = 4$, $\sigma_\eta = 0.001$.

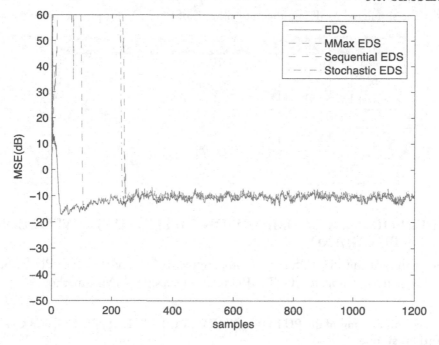

Figure 5.8: Comparison of MSE of PU EDS with EDS for white input, $N = 16$, $M = 4$, $\sigma_\eta = 0.01$.

The simulated results are obtained by taking the time average over the last 1,000 samples. The theoretical results are calculated from (5.39). The partial-update lengths are $M = 8$ and $M = 4$. We can see that the simulated results match the theoretical results.

Table 5.5: The simulated MSE and theoretical MSE of PU EDS for process noise $\sigma_\eta = 0.001$

Algorithms	Simulated MSE (dB)	Theoretical MSE (dB)
EDS (N=16)	-29.966	-30.477
MMax EDS (M=8)	-29.957	-30.470
Sequential EDS (M=8)	-29.711	-30.365
Stochastic EDS (M=8)	-29.638	-30.365
MMax EDS (M=4)	-30.269	-30.435
Sequential EDS (M=4)	-29.367	-29.929
Stochastic EDS (M=4)	-29.271	-29.938

Table 5.6: The simulated MSE and theoretical MSE of PU EDS for process noise $\sigma_\eta = 0.01$

Algorithms	Simulated MSE (dB)	Theoretical MSE (dB)
EDS (N=16)	-10.719	-11.029
MMax EDS (M=8)	-10.683	-11.029
Sequential EDS (M=8)	-10.460	-11.027
Stochastic EDS (M=8)	-10.428	-11.027
MMax EDS (M=4)	-10.567	-11.028
Sequential EDS (M=4)	-9.751	-11.022
Stochastic EDS (M=4)	-9.642	-11.022

5.6.3   PERFORMANCE COMPARISON OF THE PU EDS WITH EDS, PU RLS, RLS, PU CG, AND CG

The performance of the PU EDS is also compared with full-update EDS, PU RLS, full-update RLS, PU CG, and full-update CG. The PU method uses the MMax method.

Performance comparison of the PU EDS with EDS, PU RLS, RLS, PU CG, and CG in a time-invariant system

Figure 5.9 shows the MSE results among EDS, MMax EDS, CG, MMax CG, RLS, and MMax RLS for a time-invariant system. The same system identification model is used and the impulse response (5.40) is used. The full-update length is 16 and the partial-update length is 8. The input signal is white. We can see that the full-update EDS and MMax EDS have a convergence rate very close to the full-update RLS. The MMax EDS can converge faster than the MMax CG and MMax RLS. MMax EDS, MMax CG, and MMax RLS can achieve similar MSE to the full-update EDS, CG, and RLS at steady state.

The performance of MMax EDS, EDS, MMax RLS, RLS, MMax CG, and CG for a time-invariant system is also examined in a channel equalization system. The decision-direction channel equalizer diagram in Fig. 2.5 [43] is used. An FIR channel is used.

$$C(z) = 0.5 + 1.2z^{-1} + 1.5z^{-2} - z^{-3}. \tag{5.42}$$

The full length of the equalizer is assumed to be 30 and the PU length is 15. The input sequence is 4-QAM signal. The results are obtained by averaging 50 independent runs. Figure 5.10 illustrates the symbol-error-rate (SER) in log-scale among EDS, MMax EDS, CG, MMax CG, RLS, and MMax RLS algorithms. The SER performance of these algorithms are still related to the MSE performance shown in Figure 5.9. The full-update EDS, CG, RLS, MMax EDS, MMax CG, and MMax RLS all have similar SER performance.

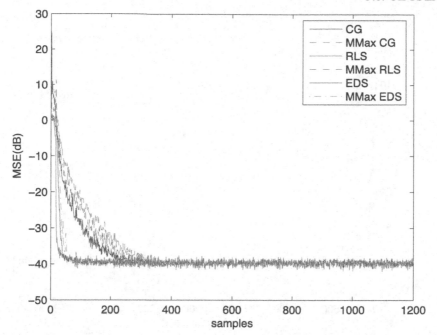

Figure 5.9: Comparison of MSE of the PU EDS with PU CG and the PU RLS.

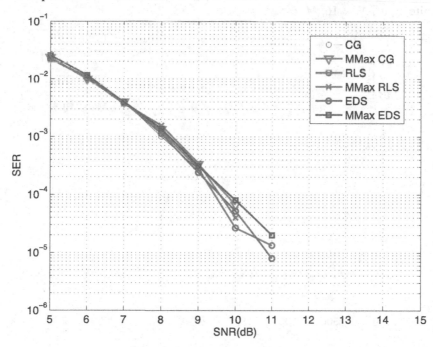

Figure 5.10: Comparison of SER of the PU EDS with the PU RLS and PU CG.

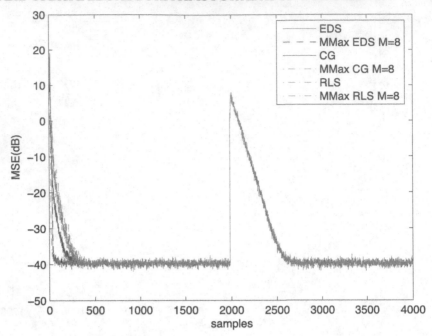

Figure 5.11: Comparison of MSE of MMax EDS with EDS, RLS, MMax RLS, CG, and MMax CG for white input, $N = 16$, $M = 8$.

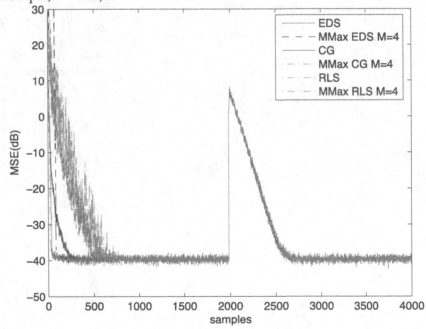

Figure 5.12: Comparison of MSE of MMax EDS with EDS, RLS, MMax RLS, CG, and MMax CG for white input, $N = 16$, $M = 4$.

Table 5.7: The computational complexities of EDS, MMax EDS, RLS, MMax RLS, CG, and MMax CG

Algorithms	Number of multiplications per symbol	Number of comparisons per symbol
EDS (N=16)	816	–
MMax EDS (M=8)	544	10
MMax EDS (M=4)	408	10
RLS (N=16)	3721	–
MMax RLS (M=8)	825	10
MMax RLS (M=4)	693	10
CG (N=16)	3003	–
MMax CG(M=8)	731	10
MMax CG (M=4)	679	10

Performance comparison of the PU EDS with EDS, PU RLS, RLS, PU CG, and CG in a time-variant system

The tracking performance of the MMax EDS is also compared with the full-update EDS, full-update RLS, MMax RLS, full-update CG, and MMax CG. The same system identification model is used. After 2,000 samples/iterations pass, the unknown system in (5.40) is changed by multiplying all coefficients by -1. Figure 5.11 and Fig. 5.12 show the MSE results among the EDS, MMax EDS, RLS, MMax RLS, CG, and MMax CG when $M = 8$ and $M = 4$, respectively. White input is used. The results show that these algorithms have a similar convergence rate after the unknown system is changed. It is also shown that the MMax EDS, MMax RLS, and MMax CG with $M = 4$ can have a similar convergence rate to the MMax EDS, MMax RLS, and MMax CG with $M = 8$ after the unknown system is changed. The partial-update length only affects the convergence rate at the beginning in this case. This is because $\widetilde{Q}(n)$ gives a less reliable estimation of $\widetilde{Q}$ at the beginning when partial-update length decreases. The EDS and MMax EDS have convergence rates very close to the full-update RLS.

Computational complexity comparison among the MMax EDS, EDS, RLS, MMax RLS, CG, and MMax CG

The SORTLINE sorting method is used for MMax EDS, MMax RLS, and MMax CG. The total number of multiplications of MMax EDS, MMax RLS, MMax CG are $N^2 + 2NM + N + 2M$, $2N^2 + 2NM + 3N + M + 1$, and $2N^2 + M^2 + 9N + M + 3$, respectively. In this case, the full-update length N is 16. The partial-update length M are 8 and 4, respectively. The detailed computational complexities of these algorithms are shown in Table 5.7. Overall, the EDS algorithms need fewer multiplications than the RLS and CG. The PU EDS algorithms can reduce the computational complexity significantly compared with full-update EDS.

5.7   CONCLUSION

In this chapter, different PU EDS algorithms are developed. Theoretical mean and mean-square performance are derived for both time-invariant and time-varying systems. The performance of different PU EDS algorithms is compared by using computer simulations. The simulated results match the theoretical results. We can conclude that the PU EDS can achieve a similar steady-state MSE as the full-update EDS while reducing the computational complexity significantly. Among different PU EDS algorithms, the MMax EDS has the best performance. The MMax EDS has a convergence rate close to the full-update EDS. The performance of the PU EDS is also compared with the PU RLS and PU CG by using MMax method. The MMax EDS can perform better than the MMax RLS and MMax CG, while having a lower computational complexity. The MMax EDS has a faster convergence rate than the MMax RLS and MMax CG. The MMax EDS, RLS, and CG can achieve the same steady-state MSE.

CHAPTER 6

Special Applications of Partial-Update Adaptive Filters

In this chapter, two special applications of PU adaptive filters are shown. One application is using PU adaptive filters to detect Global System for Mobile Communication (GSM) signals in a local GSM system using OpenBTS and Asterisk PBX. The other application is using PU adaptive filters to do image compression in a system combining hyperspectral image compression and classification.

6.1 APPLICATION IN DETECTING GSM SIGNALS IN A LOCAL GSM SYSTEM

In this section, we apply two PU adaptive filters algorithms to detect GSM signals in a local GSM system which uses OpenBTS and Asterisk PBX. A traditional GSM network is shown in Fig. 6.1 [33]. Mobile stations (MS) are connected to base transceiver stations (BTS) through a wireless air interface. Base transceiver stations are controlled by a basestation controller (BSC). The BTS reaches to a service provider's operating system through the base station controller (BSC) and switching subsystem. The switching subsystem, mobile switch center/visitor location register (MSC/VLR), is responsible for connecting voice calls and routing the Short Message Service (SMS). For commercial services, the service provider's system authenticates the mobile stations with a shared key (Ki) and an international mobile subscriber identity (IMSI) at the initial stage of mobile connection. For the GSM (global system for mobile communications) MS, the Ki and the IMSI information is stored in a subscriber identity module (SIM) card. During the initial connection, a random key is created and used for the rest of the communication.

Nowadays, the Open Base Transceiver Station (OpenBTS) and Asterisk Private Branch Exchange (PBX) are widely used for building a local GSM network for non-commercial purposes. OpenBTS is an open source software-based GSM access point. It is used to present a GSM air interface to GSM handsets. The OpenBTS acts like the traditional GSM operator network infrastructure BTS. Asterisk software PBX is a software implementation of a telephone private branch exchange (PBX) and it is used to connect calls as a MSC/VLR via Session Initiation Protocol (SIP) and Voice-over-IP (VoIP). Fig. 6.2 shows a local GSM system using OpenBTS and Asterisk PBX. Instead of building the complex infrastructure of a traditional GSM network, the local GSM system using OpenBTS and Asterisk PBX only needs a computer with a Linux

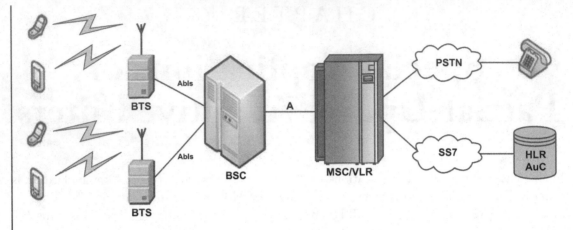

Figure 6.1: Traditional GSM network.

system and a Universal Software Radio Peripheral (USRP). OpenBTS and Asterisk are installed in a computer as normal software. The USRP is a hardware device for software radio, which can transmit and receive signals under the control of a computer and software such as GNU Radio and OpenBTS. The system controls local cellular communications. GSM cellphones can call each other, send SMS to each other, or call any SIP soft phone. The OpenBTS reads the IMSI of the MS and creates a virtual SIP including the IMSI. This virtual SIP represents a certain GSM phone. It attempts to register with Asterisk as a SIP soft phone. Once it succeeds registering with Asterisk and is provided a phone number (extension), the GSM phone can be reached by dialing that phone number [15]. Fig. 6.3 shows the functional block diagram of the local GSM system using OpenBTS, Asterisk, two USRPs, and a computer. Since there are existing commercial GSM BTS signals, a scanner is used to detect GSM signals to avoid conflicts with other GSM BTS signals. The scanner uses the first USRP, and goes through each channel (200 kHz bandwidth) in the PCS band (1930–1990 MHz). A detection algorithm is used to decide if there is a GSM signal. After all the GSM signals are found in the PCS band, out BTS uses the second USRP, and it is set up by using a channel (frequency) different from the detected signals.

The basic idea of the detection algorithm [50] in the scanner is that an adaptive line enhancer is used to detect a frequency correction burst (FCB). An FCB is a complex sine wave with frequency 67.7033 kHz. It is generated by modulating all zero bits with Gaussian minimum shift keying (GMSK). The FCB is used to do the synchronization in a GSM system and it is usually stronger than the normal burst. Using the adaptive line enhancer, we can separate narrow band sinusoid signals from noise or wideband signals. Since the FCB is a sinusoid, the periodicity is very high, and it is easier to predict by using an adaptive line enhancer than the normal burst. Figure 6.4 shows the diagram of an adaptive line enhancer. The input signal is $x(n)$, and a delay version of the input signal passes through an adaptive filter $w(n)$. The output of the adaptive filter

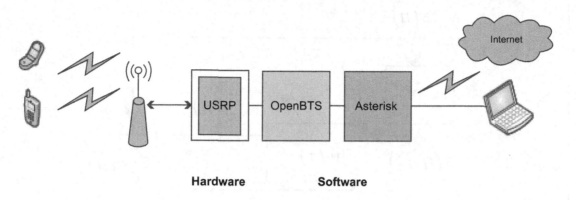

Hardware   **Software**

Figure 6.2: GSM network using OpenBTS and Asterisk PBX.

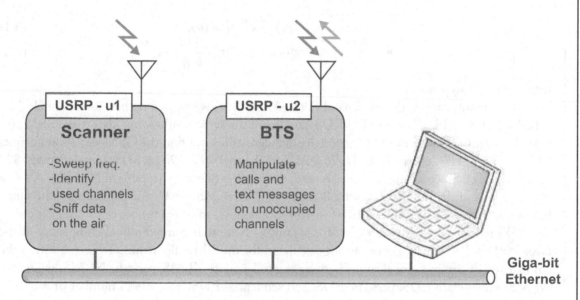

Figure 6.3: Functional-block description of local GSM system.

is $y(n)$. If the input signal is a sinusoid with some noise, then the output should be a pure sinusoid and the error between them should be very small. Adaptive filter algorithms NLMS and EDS can be used in the adaptive line enhancer. To reduce the computational complexity, the partial-update NLMS and EDS are also used.

The partial-update NLMS algorithm in an adaptive filter system is summarized as:

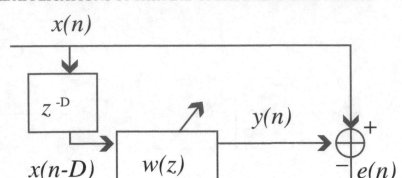

Figure 6.4: Diagram of adaptive line enhancer.

$$e(n) = d(n) - \mathbf{x}^H(n)\mathbf{w}(n), \tag{6.1}$$

$$\mathbf{w}(n+1) = \mathbf{w}(n) + \mu e^*(n)\frac{\hat{\mathbf{x}}(n)}{\|\hat{\mathbf{x}}(n)\|^2}, \tag{6.2}$$

where μ is the step size.

In our simulation, the PCS band (1930-1990MHz) is scanned. It has 300 channels. The NLMS, PU NLMS, EDS, and PU EDS can obtain the same results when detecting GSM signals. The scanner using any of the adaptive filter algorithms can find the GSM signals at channel 739 (1975.6 MHz), channel 742 (1976.2 MHz), channel 796 (1987.0 MHz), and channel 804 (1988.6 MHz). Since a complex signal is used, the number of multiplications of NLMS is $6N + 2$ per sample, and the number of multiplications of PU NLMS is $6M + 2$ per sample. The number of multiplications of EDS is $10N^2 + 10N$ per sample, and the number of multiplications of PU EDS is $6N^2 + 4NM + 6N + 4M$ per sample. Note that computational complexity of the error $e(n) = d(n) - \mathbf{x}^H(n)\mathbf{w}(n)$ is not calculated for the LMS algorithm. Table 6.1 shows the comparison of computational complexity among full-update NLMS, periodic NLMS, sequential NLMS, full-update EDS, periodic EDS, and sequential EDS. The overall number of multiplications is represented in million (M). Since the GSM signals in the PCS band are scanned, there is a total of 300 channels. We also scan 6 times for each channel to make sure the FCB can be caught. There are 10,000 data samples for each channel. We can see that the PU NLMS and PU EDS can reduce the overall number of multiplications significantly. Table 6.2 shows the overall executed time by using C language and a Linux system computer with CPU frequency 2GHz. We can see that the PU adaptive filters can save overall executed time, especially for the PU EDS algorithm. If a real-time processor with lower CPU frequency is used, the PU algorithms should be able to save more execution time.

Table 6.1: The computational complexities of the LMS, periodic LMS, sequential LMS, EDS, periodic EDS, and sequential EDS

Algorithms	Number of multiplications per sample	Overall number of multiplications
LMS(N=16)	98	1764M
periodic LMS (L=2)	98	882M
sequential LMS (N=8)	50	900M
ED S (N=16)	2720	48,960M
periodic EDS (L=2)	2720	24,480M
sequential EDS (N=8)	2176	39,168M

Table 6.2: Execution time using C and Linux

Algorithms	Time (s)
LMS(N=16)	9.50
periodic LMS (L=2)	6.11
sequential LMS (N=8)	6.01
ED S (N=16)	211
periodic EDS (L=2)	123
sequential EDS (N=8)	161

6.2 APPLICATION IN IMAGE COMPRESSION AND CLASSIFICATION

In this section, we apply two PU adaptive filter algorithms in a system combining Hyperspectral image compression and classification. A hyperspectral image contains information collected from across the electromagnetic spectrum. Spectral imaging divides the spectrum into hundreds of images/bands over a continuous spectral range. It is a three-dimensional data cube which has spatial domains (bands) and a spectral domain. A typical hyperspectral image is shown in Fig. 6.5 [27]. It can be used to discriminate the spectral features of minerals, soil, rocks, and vegetation [20]. It has been widely used in space and earth science because of its rich information about the compositional properties of surface materials. Spectral imaging applications are also used in other areas including ecology, surveillance, historical manuscript research, agriculture, etc.

Classification and compression are common operations used in image processing. Image classification is used to find special features of an image. It is usually performed on the original image, not on the image after the compression and decompression process. Because of large storage volume and spectral and spatial sample data redundancy [44], it is economical to compress hyperspectral images before transmitting, storing, and classifying them. Basic compression methods for hyperspectral imagery are transform coding based algorithms [16, 39, 47], Vector

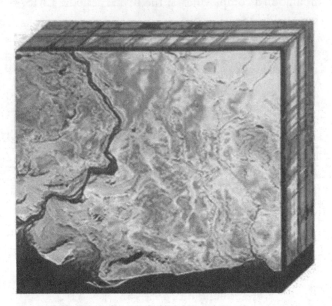

Figure 6.5: A typical hyperspectral image.

Quantization (VQ) based algorithms [37, 42], Differential Pulse Code Modulation (DPCM) [52], and Adaptive Differential Pulse Code Modulation (ADPCM) [31].

Compression and classification algorithms are conventionally performed independent of each other and performed sequentially. However, some class distinctions may be lost after a minimum distortion compression. There are very few compression/feature extraction methods which directly take classification goals (accuracy) into account in the way of feedback. Usually, compression/feature extraction is performed - with no regard to the classification objectives - before classification is performed. One of the few algorithms that accomplishes a joint optimization of feature extraction and classification is a recent iterative neural learning algorithm, called Generalized Relevance Learning Vector Quantization (GRLVQ) [23], which produces impressive results for hyperspectral data in a version further modified to specifically address high-dimensional issues [35]. In GRLVQ, classification and feature extraction are welded together with a two-way feedback loop, and classification cannot be separated and substituted by a different classifier. In this work, we propose a scheme that combines compression and classification for the hyperspectral image. The optimal compressor is chosen according to the classification results. In this scheme, the ADPCM method is used for the compression. To reduce computational complexity, partial-update (PU) adaptive filter algorithms are also used in ADPCM for the compression of hyperspectral images. The classification algorithm is the SOM-hybrid artificial neural network (ANN) (SOM-ANN). Unlike the GRLVQ, the compression method and classification method can be replaced by other methods.

A system combining hyperspectral image compression and classification is shown in Fig. 6.6. The basic idea of this system is to implement classification on both the original uncompressed image and the image after compression and decompression operations. The classification error between pre- and post-compression is fed back to the compression system in order to improve the compression system according to the optimization algorithm. In this process, the quality of classification will gradually increase, and the compression scheme will gradually improve, in order to capture important classification features. In this scheme, compression is achieved by ADPCM, which has an adaptive predictor filter and a quantizer. The filter algorithms in ADPCM are designed to minimize the mean-square or least-square cost functions. Other compression algorithms can also be used in this scheme. In this work, the ADPCM algorithm is chosen because the predictor filter can minimize the error metric, the predictor filter has different parameters to be adjusted according to the optimization algorithm, and different predictor filter algorithms can be chosen. By adjusting different parameters or choosing different predictor filter algorithms, the ADPCM algorithm can introduce flexibility into the compression and classification results. The cost function is the classification error between the uncompressed image and the compressed-decompressed image.

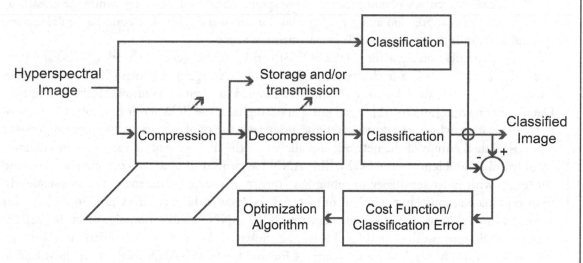

Figure 6.6: System combining compression and classification.

The optimal ADPCM predictor algorithm is chosen according to the best classification results. Different adaptive filtering algorithms for the ADPCM predictor are used in the compression system, including Least Mean Square (LMS) and Euclidean Direction Search (EDS). LMS is a very popular adaptive filtering algorithm. It has a low computational complexity but has relatively slow convergence speed and high mean-square error (MSE). The EDS algorithm has faster convergence speed and lower MSE than the LMS.

Since hyperspectral images usually have hundreds of bands and a large computer storage size, the computational complexity of compression algorithms is a big consideration. To further reduce the computational complexity, the PU LMS and PU EDS are also used for the ADPCM predictor. The partial update method aims to reduce the computational cost of the adaptive filters. Instead of updating all the $N \times 1$ coefficients, it usually only updates $M \times 1$ coefficients, where $M < N$. Two different PU methods are used. One is the periodic method, which can reduce the overall computational complexity by updating the filter coefficients periodically. The other method is stochastic PU, which can reduce the computational complexity of each sample by randomly updating parts of the filter coefficients.

The quantizer in the ADPCM is the Jayant quantizer [5]. It is an adaptive quantizer and is used to further reduce the quantization error. One-dimensional compression is applied along the spectral domain. The compression is not applied to the two-dimensional spatial domains, nor to the three-dimensional spectral and spatial domains, in order to avoid corruption of the spectra, which can be detrimental for classification accuracy. Classification of a hyperspectral image uses spectral pixels/vectors (each pixel/vector contains hundreds of bands) to do the training, therefore keeping spectra uncorrupted is more important. If a three-bit Jayant quantizer was used, the LMS was used for the predictor algorithm, and the SOM-ANN algorithm was used for classification; the classification accuracy of compression along spatial domains was 69.1%, while the classification accuracy of compression along the spectral domain was 77.3%. The compression along the spectral domain increases around 8% in classification accuracy.

The classification algorithm is the SOM-hybrid artificial neural network (ANN) architecture (SOM-ANN). It is a feed forward ANN with a self-organizing map (SOM) as a hidden layer, and a categorization learning output layer coupled to it with a Widrow-Hoff learning rule. Details are given in [26, 36, 51]. In an unsupervised phase, the SOM layer first forms its view of the data clusters, which greatly assists the accurate learning of classes in a subsequent supervised phase. The dual nature of the network also allows for the discovery of new clusters and diagnosis of learning problems in the SOM. This ANN has proved to produce both classification and clustering with great sensitivity to subtle but consistent spectral differences, and consequently, to discriminate among large numbers of material classes with high accuracy [26, 36, 51]. In this work, an improved method to implement the SOM-ANN classification algorithm is used for better statistic results. The detail for this will be shown in the section on simulations. Although we use the ADPCM method for the compression and the SOM-ANN method for classification in this work, this scheme is not limited to the ADPCM and SOM-ANN. Other compression methods and classification methods can also be used in this scheme with proper modification.

6.2.1 SIMULATIONS

The hyperspectral image we use is the Lunar Crater Volcanic Field (LCVF), taken by the AVIRIS sensor in 1994. The size of this image is 420 ×614 pixels ×194 bands. Details of this scene and of the 23 surface cover classes used in this study are given in [36, 51], which also show classification maps. One image band, at 1.02987 μm, is shown in Fig. 6.7. The whole data cube is shown in Fig. 6.8. We use this image because of the availability of previous classifications for comparison, and because we have the requisite number of labeled test samples, ≥ 100 samples per class (exclusive of the labeled training samples) to perform statistically meaningful evaluation of the classification accuracy on test data. A few of the very small classes which do not have a total of 100 samples are exceptions. Atmospheric correction and removal of excessively noisy bands leave 194 bands, and the resulting image is the source for our processing. We name this image LCVF-UNCOMP. The SOM-ANN network classified LCVF-UNCOMP into 23 classes with $\sim$ 90% overall (or weighted) accuracy (89.9% overall, $\kappa = 89.23$).

Figure 6.7: One image band of the LCVF scene, obtained at 1.02987 μm.

The comparison of classification accuracies among different predictor algorithms is shown in Table 6.3. They are also compared with the uncompressed image (LCVF-UNCOMP). The signal-to-noise ratio (SNR) is defined as the ratio of the power of the uncompressed image over the power of the error between the decompressed image and the uncompressed image. The SNR is somehow related to the classification accuracy. However, high SNR does not necessarily indicate high classification accuracy. The ADPCM predictor algorithms include LMS, periodic LMS, stochastic LMS, EDS, periodic EDS, and stochastic EDS. The periodic algorithms have a period of 4. The stochastic algorithms have a PU length of 2. The filter coefficients have a length of 8. The quantizer is the Jayant quantizer with 3 bits. AVIRIS data is represented by 12 bits per pixel/spectral channel in the original distribution. Therefore, the compression ratio is 4 : 1. The

Figure 6.8: Hyperspectral image of the LCVF scene.

main purpose of using ADPCM compression, rather than using a lossless compression method, is to adjust different parameters or use different predictor filter algorithms, and therefore it can introduce flexibility to the compression and classification results. Introducing flexibility might yield better classification results than those of the uncompressed image. In this study, we are comparing average (or un-weighted) accuracies (i.e., averages of class accuracies), in order to gain insight on how, on average, the prediction accuracies of classes are affected by the compression.

Table 6.3 shows results from two sets of experiments that we label as "Phase I" and "Phase II." These designations mark an important but unintended difference, which is the following. After the Phase I experiments, we discovered that the training and test samples used for training the classifier and for assessing its performance, respectively, had an overlap. Although this overlap was less than 10% of the entire labeled set of samples, we decided to repeat all runs with the training and test samples properly separated (with no overlap). By that time, our resources improved and we could also afford to perform three-fold jack-knives for better statistics of the classification results. This includes the variance of the classification accuracies across the three runs, which provides a measure of reliability. We label these experiments (with "clean training and test sample sets;" and three-fold jack-kniving) "Phase II." In other words, Phase I means the average of class accuracies from a single experiment, and Phase II means the average of class accuracies averaged over three jack-knife runs. In Phase I, each classification was performed only once (for reasons of limited resources), hence no averages and variances are shown. Phase I and Phase II results,

Table 6.3: The comparison of classification accuracies among different predictor algorithms

Algorithms	Average of class accuracies, from single experiment. Phase I (%)	Averages of class accuracies averaged over three jack-knives. Phase II ("clean training and test sample sets") Mean ± Std (%)	SNR (dB)	Number of multiplications per-pixel per-band	Overall number of multiplications
LCVF-UNCOMP	86.1	93.1 ± 0.46	–	–	
EDS	69.7	81.40 ± 0.40	32.937	406	20,311,660,320
Periodic EDS	74.7	82.07 ± 0.64	32.9606	103.5	5,177,972,520
Stochastic EDS	71.8	81.47 ± 0.75	32.967	102	5,102,929,440
LMS	77.3	78.37 ± 0.90	33.0656	34	1,700,976,480
Periodic LMS	74.9	81.67 ± 2.11	30.1327	20.5	1,025,588,760
Stochastic LMS	67.9	79.80 ± 1.65	30.1407	24	1,200,689,280

of course, cannot be compared with each other. However, comparisons within phases are still valuable. The classification accuracies of the uncompressed data are 86.1% for Phase I and 93.1 ± 0.46% for Phase II. The number of multiplications are also shown for each adaptive algorithm.

From Table 6.3 we can see that the classification accuracies are higher in Phase II, and the classification accuracy is not necessarily determined by the SNR. The LMS algorithm has the highest classification accuracy among all the algorithms in Phase I. The periodic EDS has the highest classification accuracy in Phase II. However, the periodic LMS and stochastic LMS have the lowest computational complexity. In Phase II, we can also see that the PU LMS and PU EDS algorithms have higher classification accuracy than the full-update LMS and EDS. It is possible that the PU adaptive filter in ADPCM can perform better than the full-update adaptive-filter while reducing the computational complexity significantly. For example, the EDS algorithm has 81.40 ± 0.40% classification accuracy in Phase II and its total number of multiplications for the whole data set is 20,311,660,320, while the periodic EDS algorithm has 82.07 ± 0.64% classification accuracy in Phase II and its total number of multiplications is 5,177,972,520. Note that the number of multiplications in the table is a per-pixel per-band number. To calculate the total number of multiplications, a constant number $420 \times 614 \times 194 = 50,028,720$, which is the number of spatial image pixels times the number of image bands, needs to be multiplied by the per-pixel per-band number of multiplications. From Table 6.3 we can see that the SNR of the LMS is higher than the SNR of the periodic LMS and stochastic LMS, however the classification accuracy of the LMS is lower than that of the periodic LMS and stochastic LMS in Phase II. This shows that high SNR does not necessarily indicate high classification accuracy.

Moreover, the classification accuracy of each class is different when using different predictor algorithms. Table 6.4 shows details of the Phase II classification error for each class when using different predictor algorithms. The results were obtained by averaging three jack-knife runs. The classification error of each class is defined by the number of misclassified samples over the total number of each class. The rows in red represent the classes with high classification errors, include classes F, I, M, V, and W. These classes are classes with few training samples and/or classes with high spectral variance. They are easily misclassified into similar classes because small differences are lost by applying the compression operation. By using different predictor filter algorithms in the ADPCM compression algorithm, it can introduce flexibility to the compression and classification results. For this LCVF data set, the periodic EDS gives the highest average classification accuracy. The periodic LMS gives the lowest classification error for classes F and I. The periodic EDS gives the lowest classification error for classes M and V. The stochastic EDS gives the lowest classification error for class W.

Table 6.4: The classification error, in units of (% misclassified samples)/100, of each class by using different predictor algorithms

Class	EDS	periodic EDS	stochastic EDS	LMS	periodic LMS	stochastic LMS
A	0.045977	0.070115	0.041379	0.041379	0.033333	0.088506
B	0.241176	0.150980	0.170588	0.211765	0.127451	0.137255
C	0.105159	0.138889	0.121032	0.073413	0.158730	0.115079
D	0.016260	0.040650	0.058072	0.042973	0.029036	0.022067
E	0.055331	0.044534	0.041835	0.028340	0.052632	0.022942
F	0.500000	0.520833	0.541667	0.531250	0.406250	0.500000
G	0.242424	0.291667	0.208333	0.250000	0.246212	0.287879
H	0.207937	0.214286	0.334920	0.315873	0.236508	0.207937
I	0.508333	0.620833	0.545833	0.529167	0.483333	0.566667
J	0.114286	0.079365	0.090476	0.050794	0.077778	0.071429
K	0.156433	0.078947	0.093567	0.102339	0.097953	0.154971
L	0.215100	0.272080	0.193732	0.286325	0.145299	0.440171
M	0.456349	0.361111	0.466270	0.736111	0.591270	0.517857
N	0.066138	0.035714	0.033069	0.058201	0.050264	0.048942
O	0.109557	0.125874	0.111888	0.144522	0.165501	0.110723
P	0.073935	0.066416	0.096491	0.090226	0.083960	0.060150
Q	0.083333	0.060606	0.064394	0.113636	0.068182	0.034091
R	0.075000	0.075000	0.041667	0.141667	0.108333	0.133333
S	0.083333	0.069444	0.104167	0.076389	0.083333	0.097222
T	0.044872	0.044872	0.076923	0.051282	0.051282	0.064103
U	0.079365	0.057540	0.063492	0.115079	0.059524	0.043651
V	0.458333	0.308333	0.433333	0.491667	0.375000	0.341667
W	0.335378	0.404908	0.325153	0.494888	0.484663	0.576687

6.3 CONCLUSION

In this chapter, the PU adaptive filters are applied in two special applications including detecting GSM signals in a local GSM system using OpenBTS and Asterisk PBX, and doing image compression in a system combining hyperspectral image compression and classification. We can see that the PU algorithms can achieve similar performance as the full-update algorithms, and sometimes perform better than the full-update algorithms.

Bibliography

[1] S. Attallah and S. W. Liaw, "Analysis of DCTLMS algorithm with a selective coefficient updating," *IEEE Trans. Circuits and Systems II: Analog and Digital Signal Processing*, vol. 48, no. 6, pp. 628-632, June 2001. DOI: 10.1109/TCSII.2005.855042. 9

[2] S. Attallah, "The wavelet transform-domain LMS adaptive filter with partial subband-coefficient updating," *IEEE Trans. Circuits and Systems II: Express Briefs*, vol. 53, no. 1, pp. 8-12, Jan. 2006. DOI: 10.1109/TCSII.2005.855042. 9

[3] B. G. Agee, "The least-squares CMA: A new technique for rapid correction of constant modulus signals," in *Proc. International Conference Acoustics, Speech, and Signal Processing*, Apr. 1986, vol. 11, pp. 953-956. DOI: 10.1109/ICASSP.1986.1168852. 15

[4] B. G. Agee, "The property restoral approach to blind adaptive signal extraction," PhD dissertation, Department of Electrical Engineering and Computer Science, University of California, Davis, CA, June, 1989.

[5] T. Bose, *Digital Signal and Image Processing*. New York: Wiley, 2004. 54, 68, 92

[6] A. Carini and G. L. Sicuranza, "Analysis of transient and steady-state behavior of a multichannel filtered-x partial-error Affine Projection algorithm," *EURASIP Journal on Audio, Speech, and Music Processing*, vol. 2007 ID. 31314. DOI: 10.1155/2007/31314. 10

[7] P. S. Chang and A. N. Willson, "Analysis of conjugate gradient algorithms for adaptive filtering," *IEEE Trans. Signal Processing*, vol. 48, no. 2, pp. 409-418, Feb. 2000. DOI: 10.1109/78.823968. 47, 50, 69

[8] G. Deng, "Partial update and sparse adaptive filters," *IET Signal Processing*, vol. 1, no. 1, pp. 9-17, 2007. DOI: 10.1049/iet-spr:20060128. 9

[9] P. S. R. Diniz, G. O. Pinto, and A. Hjorungnes, "Data selective partial-update affine projection algorithm," in *Proc. International Conference Acoustics, Speech, and Signal Processing*, Apr. 2008, pp. 3833-3836. DOI: 10.1109/ICASSP.2008.4518489. 10

[10] K. Doğançay and O. Tanrikulu, "Normalised constant modulus algorithm with selective partial updates," in *Proc. International Conference Acoustics, Speech, and Signal Processing*, May 2001, vol. 4, pp. 2181-2184. DOI: 10.1109/ICASSP.2001.940427. 10

[11] K. Doğançay, *Partial-Update Adaptive Signal Processing*. Academic Press, 2008. 1, 9, 10, 11, 12, 16, 49, 58

[12] S. C. Douglas, "Analysis and implementation of the max-NLMS adaptive filter," in *Proc. Twenty-Ninth Asilomar Conference on Signals, Systems and Computers*, 1995, vol. 1, pp. 659-663. DOI: 10.1109/ACSSC.1995.540631. 9, 10

[13] S. C. Douglas, "Adaptive filters employing partial updates," *IEEE Trans. Circuits and Systems II: Analog and Digital Signal Processing*, vol. 44, no. 3, pp. 209-216, Mar. 1997. DOI: 10.1109/82.558455. 9, 10

[14] D. L. Duttweiler, "Proportionate normalized least-mean-squares adaptation in echo cancelers," *IEEE Trans. Speech, and Audio Processing*, vol. 8, no. 5, pp. 508-518, Sept. 2000. DOI: 10.1109/89.861368. 9

[15] B. Ferguson, [Online]. "Communication between GSM devices and the PSTN via integration of Google Voice and OpenBTS with the Asterisk PBX," 2011. 86

[16] J. E. Fowler and J. T. Rucker, "3D wavelet-based compression of hyperspectral imagery," *Hyperspectral Data Exploitation: Theory and Applications*, C.-I. Chang, Ed., pp. 379-407, chapter 14, John Wiley & Sons, Hoboken, NJ, USA, 2007. 89

[17] H. Furukawa, Y. Kamio, and H. Sasaoka, "Co-channel interference reduction method using CMA adaptive array antenna," in *Proc. 7th IEEE International Symposium Personal, Indoor, Mobile Radio Communications*, Oct. 1996, vol. 2, pp. 512-516. DOI: 10.1109/PIMRC.1996.567447.

[18] D. N. Godard, "Self-recovering equalization and carrier tracking in two-dimensional data communication systems," *IEEE Trans. Commun.*, vol. 28, no. 11, pp. 1867-1875, Nov. 1980. DOI: 10.1109/TCOM.1980.1094608. 14

[19] M. Godavarti and A. O. Hero III, "Partial update LMS algorithms," *IEEE Trans. Signal Processing*, vol. 53, no. 7, pp. 2382-2399, 2005. DOI: 10.1109/TSP.2005.849167. 9, 10

[20] R. O. Green, "Summaries of the 6th annual JPL airborne geoscience workshop," in *Proc. AVIRIS Workshop*, Mar. 1996, vol. 1. 89

[21] S. Haykin, *Adaptive Filter Theory*. 3th ed. New Jersey: Prentice-Hall, 1996. 1

[22] S. Haykin, *Adaptive Filter Theory*, 4th ed. New Jersey: Prentice Hall, 2002. 13, 14, 43, 52, 53, 71, 72

[23] B. Hammer, and Th. Villmann, "Generalized relevance learning vector quantization," *Neural Networks*, vol. 15, pp. 1059-1068, 2002. DOI: 10.1016/S0893-6080(02)00079-5. 90

[24] K. Hilal and P. Duhamel "A convergence study of the constant modulus algorithm leading to a normalized-CMA and block-normalized-CMA," in *Proc. EUSIPCO*, Aug. 1992, pp. 135-138. 14

[25] M. L. Honig, U. Madhow, and S. Verdu, "Blind adaptive interference suppression for the near-far resistant CDMA," in *Proc. GLOBECOM*, 1994, vol. 1, pp. 379-384. DOI: 10.1109/78.552211.

[26] E. S. Howell, E. Merényi, and L. A. Lebofsky, "Classification of asteroid spectra using a neural network," *Geophysics Research*, vol. 99, no. 10, pp. 847-865, 1994. DOI: 10.1029/93JE03575. 92

[27] [Online]. Available: http://rst.gsfc.nasa.gov/Intro/Part2_24.html 89

[28] C. R. Johnson, Jr., P. Schnitter, T. Endres, J. Behm, D. Brown and R. A. Casas, "Blind equalization using the constant modulus criterion: A review," *Proc. IEEE*, vol. 86, no. 10, pp. 1927-1950, Oct. 1998. DOI: 10.1109/5.720246.

[29] A. W. H. Khong and P. A. Naylor, "Selective-tap adaptive filtering with performance analysis for identification of time-varying systems," *IEEE Trans. Audio, Speech, and Language Processing*, vol. 15, no. 5, pp. 1681-1695, July 2007. DOI: 10.1109/TASL.2007.896671. 9, 10

[30] A. W. H. Khong, P. A. Naylor, and J. Benesty, "A low delay and fast converging improved proportionate algorithm for sparse system identification," *EURASIP Journal on Audio, Speech, and Music Processing*, vol. 2007, ID. 84376. DOI: 10.1155/2007/84376. 9

[31] M. Larsen, T. Bose, and A. Venkatachalam, "Hyperspectral image restoration and coding," in *Proc. Thirty-Sixth Asilomar Conference on Signals, Systems and Computers*, Nov. 2002, vol.2, pp. 1740-1744. DOI: 10.1109/ACSSC.2002.1197073. 90

[32] W. Lee, B. R. Vojcic, and R. L. Pickholtz, "Constant modulus algorithm for blind multiuser detection," in *Proc. IEEE 4th International Symposium on Spread Spectrum Techniques and Applications Proceedings*, Sept. 1996, vol. 3, pp. 1262-1266. DOI: 10.1109/ISSSTA.1996.563508.

[33] A. Loula, [Online]. "OpenBTS Installation and Configuration Guide v0.1," 2009. 85

[34] K. Mayyas, "Performance analysis of the deficient length LMS adaptive algorithm," *IEEE Trans. Signal Processing*, vol. 53, no. 8, pp. 2727-2734, Aug. 2005. DOI: 10.1109/TSP.2005.850347. 9, 54, 73

[35] M. J. Mendenhall, and E. Merényi, "Relevance-based feature extraction for hyperspectral images," *IEEE Trans. Neural Networks* vol. 19, no. 4, pp 658-672, April 2008. DOI: 10.1109/TNN.2007.914156. 90

[36] E. Merényi, W.H. Farrand, J.V. Taranik, and T.B. Minor, "Classification of hyperspectral imagery with neural networks: comparison to conventional tools," *Machine Learning Reports 04/2011* T. Villmann and F.-M. Schleif, Ed., vol. 5, no. 4, pp 1-15, April 2011, ISSN:1865-3960 `http://www.techfak.uni-bielefeld.de/~fschleif/mlr/mlr_04_2011.pdf`. Also submitted to EURASIP Journal on Advances in Signal Processing, special issue on Neural Networks for Remotely Sensed Data Interpretation, 2011. 92, 93

[37] G. Motta, F. Rizzo, and J. A. Storer, "Compression of hyperspectral imagery," in *Proc. Data Compression Conference*, Mar. 2003, pp. 25-27. DOI: 10.1109/DCC.2003.1194024. 90

[38] P. A. Naylor and A. W. H. Khong, "Affine projection and recursive least squares adaptive filters employing partial updates," in *Proc. Thirty-Eighth Asilomar Conference on Signal, System & Computers*, Nov. 2004, vol. 1, pp. 950-954. DOI: 10.1109/ACSSC.2004.1399279. 10

[39] B. Penna, T. Tillo, E. Magli, and G. Olmo, "Transform coding techniques for lossy hyperspectral data compression," *IEEE Trans. Geosci. Remote Sens.*, vol. 45, no. 5, pp. 1408-1421, May 2007. DOI: 10.1109/TGRS.2007.894565. 89

[40] M. R. Petraglia, and D. B. Haddad, "New adaptive algorithms for identification of sparse impulse responses - Analysis and comparisons," in *Proc. ISWCS*, 2010, pp. 384-388. DOI: 10.1109/ISWCS.2010.5624301. 9

[41] [Online]. Available: `http://www.calstatela.edu/centers/berp/docs/Wireless_Channel_Models.pdf`

[42] M. J. Ryan and J. F. Arnold, "The lossless compression of AVIRIS images by vector quantization," *IEEE Trans. Geosci. Remote Sens.*, vol. 35, no. 3, May 1997. DOI: 10.1109/36.581964. 90

[43] A. H. Sayed, *Fundamentals of Adaptive Filtering*. New York: Wiley, 2003. 9, 21, 35, 58, 80

[44] G. Shaw and D. Manolakis, "Signal processing for hyperspectral image exploitation," *Signal Processing Mag.*, vol. 19, no. 1, Jan. 2002. DOI: 10.1109/79.974715. 89

[45] A. Sugiyama, H. Sato, A. Hirano, and S. Ikeda, "A fast convergence algorithm for adaptive FIR filters under computational constraint for adaptive tap-position control," *IEEE Trans. Circuits and Systems II: Analog and Digital Signal Processing*, vol. 43, no. 9, pp. 629-636, Sept. 1996. DOI: 10.1109/82.536759. 9

[46] A. Sugiyama, S. Ikeda, and A. Hirano, "A fast convergence algorithm for sparse-tap adaptive FIR filters identifying an unknown number of dispersive regions," *IEEE Trans. Signal Processing*, vol. 50, no. 12, pp. 3008-3017, Dec. 2002. DOI: 10.1109/TSP.2002.805255. 9

[47] X. Tang and W. A. Pearlman, "Lossy-to-lossless block-based compression of hyperspectral volumetric data," in *Proc. IEEE Int. Conf. Image Process.*, 2004, pp. 3283-3286. DOI: 10.1109/ICIP.2006.312756. 89

[48] J. R. Treichler and B. G. Agee, "A new approach to multipath correction of constant modulus signals," *IEEE Trans. Acoustics, Speech, and Signal Processing*, vol. 31, no. 2, pp. 459-472, Apr. 1983. DOI: 10.1109/TASSP.1983.1164062. 13, 14

[49] J. R. Treichler and M. G. Larimore, "The tone capture properties of CMA-based interference suppressors," *IEEE Trans. Acoustics, Speech, and Signal Processing*, vol. 33, no. 4, pp. 946-958, Aug. 1985. DOI: 10.1109/TASSP.1985.1164640.

[50] G. N. Varma, U. Sahu, and G. P. Charan, "Robust frequency burst detection algorithm for GSM/GRPS," in *Proc. IEEE Vehicular Technology Conference*, Sept. 2004, vol. 6, pp. 3843-3846. DOI: 10.1109/VETECF.2004.1404796. 86

[51] T. Villmann, E. Merényi, and B. Hammer, "Neural maps in remote sensing image analysis," *Neural Networks*, vol. 16, pp. 389-403, 2003. DOI: 10.1016/S0893-6080(03)00021-2. 92, 93

[52] H. Wang and K. Sayood, "Lossless predictive compression of hyperspectral images," *Hyperspectral Data Compression*, pp. 35-56. Springer, 2006. DOI: 10.1007/0-387-28600-4_2. 90

[53] S. Werner, J. A. Apolinário, Jr., and P. S. R. Diniz, "Set-membership proportionate Affine Projection algorithms," *EURASIP Journal on Audio, Speech, and Music Processing*, vol. 2007, ID. 34242. DOI: 10.1155/2007/34242. 10

[54] J. Wu, M. Doroslovacki, "A mean Convergence analysis for partial update NLMS algorithms," in *Proc. CISS*, Mar. 2007, pp. 31-34. DOI: 10.1109/CISS.2007.4298268. 9

[55] G. F. Xu and T. Bose, "Analysis of the Euclidean direction search adaptive algorithm," in *Proc. International Conference on Acoustics, Speech and Signal Processing*, pp. 1689-1692, Apr. 1998. DOI: 10.1109/ICASSP.1998.681781. 67

[56] G. F. Xu, "Fast algorithms for digital filtering: Theory and applications," Ph.D. dissertation, University of Colorado, 1999.

[57] Z. K. Zhang, T. Bose, L. Xiao, and R. Thamvichai, "Performance analysis of the deficient length EDS adaptive algorithms," in *Proc. APCCAS*, Dec. 2006, pp. 222-226. DOI: 10.1109/TSP.2005.850347. 54, 68, 73

Authors' Biographies

BEI XIE

Bei Xie received a Ph.D. in electrical engineering from Virginia Polytechnic Institute and State University in 2012. Her interests include signal processing and communications.

TAMAL BOSE

Dr. Tamal Bose serves as Professor and Department Head of Electrical and Computer Engineering at the University of Arizona. He is also the Director of a multi-university NSF Center called the Broadband Wireless Access & Applications Center (BWAC).

Dr. Bose's research interests include signal classification for cognitive radios, channel equalization, adaptive filtering algorithms, and nonlinear effects in digital filters. He is author of the text *Digital Signal and Image Processing*, John Wiley, 2004, and coauthor of *Basic Simulation Models of Phase Tracking Devices Using MATLAB*, Morgan & Claypool Publishers, 2010.